First Workshop on Natural Language Processing for Medical and Healthcare Fields 2013

Held at the 6th International Joint Conference on Natural Language Processing (IJCNLP 2013)

Nagoya, Japan
14-18 October 2013

ISBN: 978-1-62993-546-1

Printed from e-media with permission by:

Curran Associates, Inc.
57 Morehouse Lane
Red Hook, NY 12571

Some format issues inherent in the e-media version may also appear in this print version.

Printed by Curran Associates, Inc. (2014)

Association for Computational Linguistics
209 N. Eighth Street
Stroudsburg, Pennsylvania 18360

Phone: 1-570-476-8006
Fax: 1-570-476-0860

acl@aclweb.org

Additional copies of this publication are available from:

Curran Associates, Inc.
57 Morehouse Lane
Red Hook, NY 12571 USA
Phone: 845-758-0400
Fax: 845-758-2634
Email: curran@proceedings.com
Web: www.proceedings.com

We wish to thank our sponsors and supporters!

Platinum Sponsors

www.anlp.jp

Silver Sponsors

www.google.com

Bronze Sponsors

www.rakuten.com

Supporters

Nagoya Convention & Visitors Bureau

We wish to thank our organizers!

Organizers

AFNLP

Asian Federation of Natural Language Processing (AFNLP)

Master technology, create technology

TOYOHASHI UNIVERSITY OF TECHNOLOGY

Toyohashi University of Technology

Preface

Welcome to the Workshop on Natural Language Processing for Medical and Healthcare Fields. We received 9 submissions. After one withdrawal, we chose to accept four long papers and four short papers, giving an long paper acceptance rate of 50%.

Medical records are increasingly written on electronic media instead of on paper, which has radically increased the importance of information processing techniques in medical fields. Nevertheless, the state of usage of information and communication technologies in medical fields is said to 10 years behind that in other fields. By processing large amounts of medical records and obtaining knowledge from them, great potential exist in assisting more precise and timely treatments. Such assistance can save lives and provide better quality of life.

Our goal is the promotion and support of implementation of practical tools and systems in the medical industry, which can support medical decisions and treatment by physicians and medical staff. A short-term objective of this pilot task is to evaluate basic techniques of information extraction in medical fields, but the long-term objective is to offer a forum for achieving the goal with a community-based approach. We aim to gather people who are interested in this issue. Then we intend to facilitate their communication and discussion to clarify issues to be solved, while defining the necessary elemental technologies.

Finally, we would like to express our gratitude to the following people who helped us. Thank you.
The Workshop on Natural Language Processing for Medical and Healthcare Fields Program Committee

Organizers:

Eiji Aramaki, Kyoto University, Japan.
Mizuki Morita, The University of Tokyo, Japan.

Program Committee:

Yuki Arase, MicroSoft Research Asia, China.
Eric Chang, MicroSoft Research Asia, China.
Nigel H. Collier, National Institute of Informatics, Japan.
Son Doan, University of California San Diego, USA.
Yoshinobu Kano, JST, Japan.
Mai Miyabe, The University of Tokyo, Japan.
Takashi Okumura, National Institute of Public Health, Japan.
Hoifung Poon, MicroSoft Research, USA.
Ozlem Uzuner, SUNY Albany, USA.

Table of Contents

Incorporating Knowledge Resources to Enhance Medical Information Extraction
Yasuhide Miura, Tomoko Ohkuma, Hiroshi Masuichi, Emiko Yamada Shinohara, Eiji Aramaki and Kazuhiko Ohe 1

Clinical Vocabulary and Clinical Finding Concepts in Medical Literature
Takashi Okumura, Eiji Aramaki and Yuka Tateisi 7

Developing ML-based Systems to Extract Medical Information from Japanese Medical History Summaries
Shohei Higashiyama, Kazuhiro Seki and Kuniaki Uehara 14

Towards High-Reliability Speech Translation in the Medical Domain
Graham Neubig, Sakriani Sakti, Tomoki Toda, Satoshi Nakamura, Yuji Matsumoto, Ryosuke Isotani and Yukichi Ikeda 22

Finding Every Medical Terms by Life Science Dictionary for MedNLP
Shuj Kaneko, Nobuyuki Fujita and Hiroshi Ohtake 30

Proper and Efficient Treatment of Anaphora and Long-Distance Dependency in Context-Free Grammar: An Experiment with Medical Text
Wailok Tam, Koiti Hasida, Yusuke Matsubara, Eiji Aramaki, Mai Miyabe, Motoyuki Takaai and Hirosi Uozaki 33

A Comparison of Rule-Based and Machine Learning Methods for Medical Information Extraction
Osamu Imaichi, Toshihiko Yanase and Yoshiki Niwa 38

The Electronic Health Record as a Clinical Study Information Hub
Naoto Kume, Kazuya Okamoto, Tomohiro Kuroda and Hiroyuki Yoshihara 43

Conference Program

Oct 18, 2013

Session: Long Paper

Incorporating Knowledge Resources to Enhance Medical Information Extraction
Yasuhide Miura, Tomoko Ohkuma, Hiroshi Masuichi, Emiko Yamada Shinohara, Eiji Aramaki and Kazuhiko Ohe

Clinical Vocabulary and Clinical Finding Concepts in Medical Literature
Takashi Okumura, Eiji Aramaki and Yuka Tateisi

Developing ML-based Systems to Extract Medical Information from Japanese Medical History Summaries
Shohei Higashiyama, Kazuhiro Seki and Kuniaki Uehara

Towards High-Reliability Speech Translation in the Medical Domain
Graham Neubig, Sakriani Sakti, Tomoki Toda, Satoshi Nakamura, Yuji Matsumoto, Ryosuke Isotani and Yukichi Ikeda

Session: Short Paper

Finding Every Medical Terms by Life Science Dictionary for MedNLP
Shuj Kaneko, Nobuyuki Fujita and Hiroshi Ohtake

Proper and Efficient Treatment of Anaphora and Long-Distance Dependency in Context-Free Grammar: An Experiment with Medical Text
Wailok Tam, Koiti Hasida, Yusuke Matsubara, Eiji Aramaki, Mai Miyabe, Motoyuki Takaai and Hirosi Uozaki

A Comparison of Rule-Based and Machine Learning Methods for Medical Information Extraction
Osamu Imaichi, Toshihiko Yanase and Yoshiki Niwa

The Electronic Health Record as a Clinical Study Information Hub
Naoto Kume, Kazuya Okamoto, Tomohiro Kuroda and Hiroyuki Yoshihara

Incorporating Knowledge Resources to Enhance Medical Information Extraction

Yasuhide Miura
Fuji Xerox Co., Ltd., Japan
yasuhide.miura@fujixerox.co.jp

Tomoko Ohkuma
Fuji Xerox Co., Ltd., Japan
ohkuma.tomoko@fujixerox.co.jp

Hiroshi Masuichi
Fuji Xerox Co., Ltd., Japan
hiroshi.masuichi@fujixerox.co.jp

Emiko Yamada Shinohara
The University of Tokyo, Japan
emiko-tky@umin.net

Eiji Aramaki
Kyoto University, Japan
JST PRESTO, Japan
eiji.aramaki@gmail.com

Kazuhiko Ohe
The University of Tokyo Hospital, Japan
The University of Tokyo, Japan
kohe@hcc.h.u-tokyo.ac.jp

Abstract

This paper describes a method to extract medical information from texts. The method targets to extract complaints and diagnoses from electronic health record texts. Complaints and diagnoses are fundamental information and can be used for more complex medical tasks. The method utilizes several medical knowledge resources to enhance the performance of extraction. With an evaluation using NTCIR-10 MedNLP data, our method marked 86.53 in F_1 score with a cross validation. The score is comparable to top scoring teams in NTCIR-10 MedNLP task. The approach taken to incorporate knowledge resources has a high generality. It is not restricted to the resources presented in this paper and can be applied to various other resources.

1 Introduction

Spread of electronic health record (EHR) brought a large amount of unstructured medical data that can be processed electronically. The data include valuable information about patients health. An automatic extraction of medical information from them is beneficial since manual analyses of them by medical experts are often difficult because of their quantity. This paper describes a method that enables such automatic extraction.

Various kinds of information have been targeted for an extraction from EHR texts. NLP shared tasks of Informatics for Integrating Biology and the Bedside (i2b2)[1] designed challenges to extract kinds of medical information such as: smoking status, obesity, medication, medical problem, medical test, and treatment. Medical Records tracks of Text Retrieval Conference (TREC)[2] modeled an extraction of cohorts that are effective for medical researches. Natural Language Processing (MedNLP) task of NTCIR[3] aimed to extract patient complaints and diagnoses from EHR texts. The method described in this paper targets to extract complaints and diagnoses, the same kind of information that NTCIR-10 MedNLP intended. Complaints and diagnoses are fundamental information and can be useful for complex medical tasks. Example of such tasks are: an assignment of disease codes and a detection of adverse effects in medications. Figure 1 shows an example of a diagnosis description in an EHR text.

EN No <c modality="negation">edema</c> on the front shin bone part.

JA 前脛骨部に<c modality="negation">浮腫</c>なし。

Figure 1: An example of a diagnosis description in an EHR text of NTCIR-10 MedNLP data. In NTCIR-10 MedNLP, c tag is used to denote a complaint or a diagnosis.

The extraction of complaints and diagnoses is known to achieve a moderate performance (78.86 in F_1 score) by applying a simple conditional random field (CRF) based named entity recognition (NER) method (Imachi et al., 2013). Our method utilizes medical knowledge to a CRF based NER method to enhance an extraction performance. Our contribution in this paper is that we show a substantial increase in the extraction performance of complaints and diagnoses by incorporating several medical knowledge resources. The paper

[1]https://www.i2b2.org/

[2]http://trec.nist.gov/

[3]http://ntcir.nii.ac.jp/

also discusses the detailed effects of the individual knowledge resources and the generality of the method.

The outline of this paper is as follows. Section 2 explains the detail of the method. Section 3 describes an experiment we performed for an evaluation. Section 4 notes the related works. Section 5 discusses the result of the experiment and the generality of the method. Section 6 concludes the paper.

2 Method

A method that we prepared for a medical information extraction is basically a machine learning based named entity recognizer. The method assumes that information to be extracted can be expressed as named entities. NER can be interpreted as a sequential labeling problem. We utilized linear-chain CRF (Lafferty et al., 2001), one of widely used methods to handle the problem, with character-level node. Character-level processing is chosen since Japanese text is unsegmented text and a character-level NER is known to achieve the state-of-the-art accuracy (Asahara and Matsumoto, 2003).

NER is known as a knowledge-intensive task and the use of external knowledge often boosts the performance of it (Ratinov and Roth, 2009). Various knowledge resources (e.g. dictionary, terminology, ontology) are available in medical fields. We decided to exploit three publicly available medical terminologies, MedDRA/J[4] (Brown et al., 1999), MEDIS Byomei Master[5] (Medical Information System Development Center, 2012), and MEDIS Shojo Shoken Master ⟨Shintai Shoken Hen⟩[6].

Additionally to these terminologies, we also utilized information obtained from an external corpus in a medical domain. We introduced named entities that are defined on the updated version of the discharge summary corpus (DS Corpus) mentioned in Aramaki et al. (2009). DS corpus contains *symptom* named entities and *disease* named entities that are similar to complaints and diagnoses in NTCIR-10 MedNLP task. BASELINE composition of our method (detail will be described in Section 3.1) was trained on DS Corpus to realize a DS Corpus named entity recognizer.

[4]http://www.pmrj.jp/jmo/php/indexe.php

[5]http://www2.medis.or.jp/stdcd/byomei/index.html (In Japanese)

[6]http://www2.medis.or.jp/master/syoken/ (In Japanese)

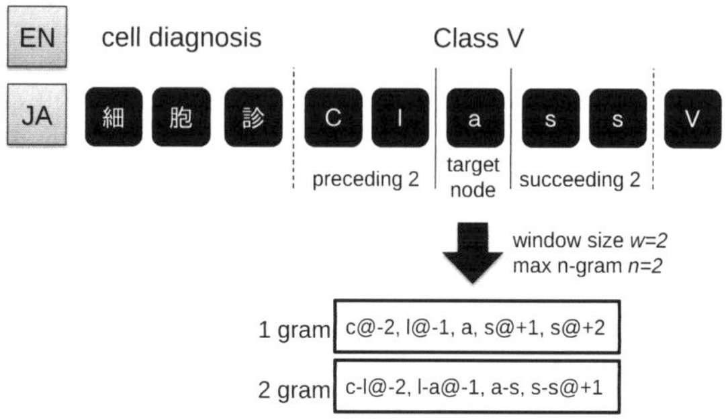

Figure 2: An example of sliding window features of "C-SURF" with window size $w = 2$ and max n-gram $n = 2$. A number following "@" represents the position from the target node.

Table 1 lists all features that are used in our method. For all features, sliding window features illustrated in figure 2 are considered. All features derive information from character, morpheme, or external knowledge. Therefore several preprocesses are done prior to the feature extraction. A morphological analysis and assignments of the resulting morphemes to character nodes are done to extract "M-*" features. A BIO-style match of the three terminologies similar to Kazama and Torisawa (2007) is applied to extract "K-MEDDRA", "K-MEDIS-BM", and "K-MEDIS-SSM" features. The DS Corpus named entities are recognized and the BIO-style matches of them are performed to extract "K-NE-SD" feature.

2.1 Implementation

This section briefly describes the method in an implementation perspective. Figure 3 portraits the architecture of the method.

Text Normalization Module

Three simple text normalization processes are applied to an input text as a first step. Firstly, a Unicode normalization in form NFKC[7] is applied. Secondly, all upper case characters are converted to lower case ones based on the definition of Unicode Standard version 4.0. Thirdly, all half-width characters are converted to full-width characters using ICU[8].

Character Analysis Module

Unicode blocks that the characters of a text belong to are extracted as character types.

[7]http://unicode.org/reports/tr15/

[8]http://site.icu-project.org/

Feature	Description
C-SURF	The surface form of a character.
C-TYPE	The type of a character. The Unicode block[i] is used for the type category.
M-SURF	The surface form of a morpheme.
M-BASE	The base form of a morpheme.
M-POS1	The part-of-speech layer 1 of a morpheme.
M-POS2	The part-of-speech layer 2 of a morpheme.
M-POS3	The part-of-speech layer 3 of a morpheme.
M-CJ-FORM	The conjugation form of a morpheme.
M-CJ-TYPE	The conjugation type of a morpheme.
K-MEDDRA	The BIO-style matching result of a character with MedDRA/J entries.
K-MEDIS-BM	The BIO-style matching result of a character with MEDIS Byomei Master entries.
K-MEDIS-SSM	The BIO-style matching result of a character with MEDIS Shojo Shoken Master ⟨Shintai Shoken Hen⟩ entries.
K-NE-SD	The BIO-style matching result of a character with recognized DS Corpus symptom named entities and DS Corpus disease named entities.

[i] http://www.unicode.org/charts/

Table 1: The list of features used in the method.

Morphological Analysis Module

A morphological analysis is applied to a text using Kuromoji[9] with mode set to "Search". Assignments of resulting morphemes to corresponding characters are also done in this module.

External Named Entity Annotation Module

DS Corpus trained named entity recognizers are applied to a text. For each named entity recognizer, assignments of BIO-style tags to each character are also done in this module.

External Terminology Annotation Module

The entries in the three medical terminologies (MedDRA/J, MEDIS Byomei Master, and MEDIS Shojo Shoken Master ⟨Shintai Shoken Hen⟩) are matched to a text. For each terminology, assignments of BIO-style tags (e.g. "B-K-MEDIS-BM", "I-K-MEDIS-BM") to each character are also done in this module.

Feature Aggregation Module

Features are aggregated based on a feature composition and are encoded to the input format of the machine learning module. Sliding window features are set here with the parameters of window size w and max gram size n. A simple frequency based feature filtering is also available to ignore sparse features with frequency threshold t.

[9]http://www.atilika.org/

Machine Learning Module

CRF is applied to aggregated features. For the implementation of CRF, MALLET[10] is used with default parameters.

3 Experiment

3.1 Feature Compositions

We prepared six feature compositions of the method. Table 2 lists all compositions and their feature sets. BASELINE is a composition that we prepared as a baseline of the method. It only consists of the features based on character and morpheme. DSNE adds the named entity feature to BASELINE. MEDDRA and MEDIS add one terminology feature to BASELINE. MEDDIC adds all terminology features to BASELINE. FULL uses all defined features.

3.2 Evaluation Data

A performance of our method was evaluated on the training portion of NTCIR-10 MedNLP data. The data consist of 2,244 sentences with 1,922 complaint or diagnosis (*c* tag) annotations. Modality information included in some of *c* tags is not considered in this experiment. The detail of the data can be found in the overview paper of NTCIR-10 MedNLP (Morita et al., 2013).

[10]http://mallet.cs.umass.edu/

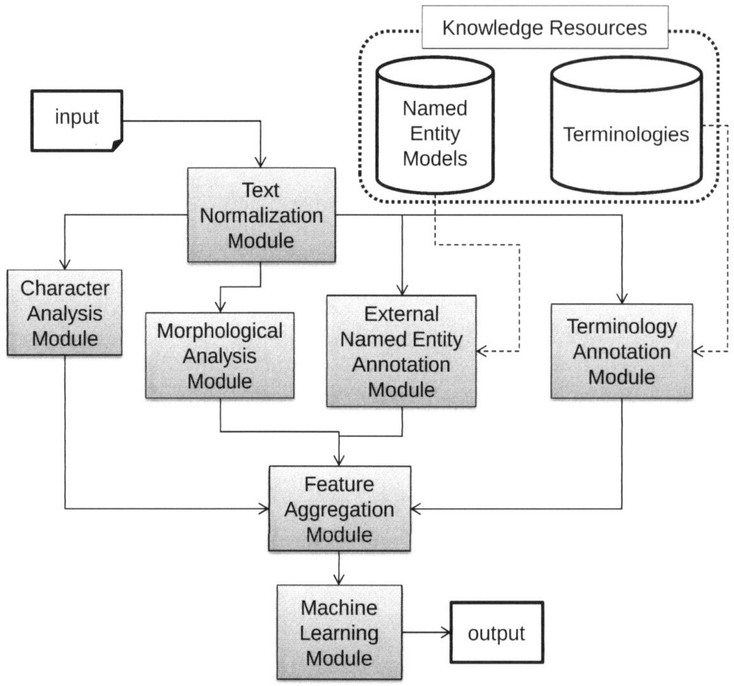

Figure 3: The architecture of the method.

Composition	Feature Sets
BASELINE	{C-SURF, C-TYPE, M-SURF, M-BASE, M-POS1, M-POS2, M-POS3, M-CJ-FORM, M-CJ-TYPE}
DSNE	BASELINE ∪ {K-NE-SD}
MEDDRA	BASELINE ∪ {K-MEDDRA}
MEDIS	BASELINE ∪ {K-MEDIS-BM, K-MEDIS-SSM}
MEDDIC	MEDDRA ∪ MEDIS
FULL	DSNE ∪ MEDDIC

Table 2: The list of feature compositions.

3.3 Extraction Performance

We measured precisions, recalls, and F_1 scores of c tag extraction as extraction performances. 5-fold cross validations were ran on six system compositions: BASELINE, DSNE, MEDDRA, MEDIS, MEDDIC, and FULL. The parameters of the feature aggregation module were set to $w = 2$, $n = 2$, and $t = 2$. Table 3 shows the micro average 5-fold cross validation values of the six compositions.

A statistical significance of two compositions was tested by a randomization test described in Noreen (1989) with iteration number set to 10,000. Statistical significances between six compositions were tested by five pairs: DSNE–BASELINE, MEDDRA–BASELINE, MEDIS–BASELINE, MEDDIC–BASELINE, and FULL–MEDDIC. Statistically significant improvements with $p \leq 0.05$ were achieved in, the recall and the F_1 score of DSNE, the precision, the recall, and the F_1 score of MEDDRA, the precision and the F_1 score of MEDIS, the precision, the recall, and the F_1 score of MEDDIC, and the recall of FULL.

Composition	Precision	Recall	F_1 score
BASELINE	87.87%	81.43%	84.53
DSNE	87.46%	84.18%	85.79
MEDDRA	88.88%	82.78%	85.72
MEDIS	89.40%	82.52%	85.82
MEDDIC	88.57%	83.45%	85.94
FULL	88.39%	84.76%	86.53

Table 3: The 5-fold cross validation results of the method. The underlined values represent statistically significant improvements.

4 Related Works

NER is well studied in the field of natural language processing. A number of design issues in NER are discussed in Ratinov and Roth (2009). This section explains NER methods that have close relationship with our method.

A character-level processing of NER is investigated in some literatures. Asahara and Matsumoto (2003) showed that a state-of-the-art Japanese

Terminology	# of Terms
MedDRA/J	922
MEDIS BM & SSM	1,041
MedDRA/J ∩ MEDIS BM & SSM	421

Table 4: The number of terms that are present in NTCIR-10 MedNLP data for each terminology. MEDIS BM & SSM is the union of the two MEDIS terminologies that we used.

NER can be realized with character level processing. Klein et al. (2003) demonstrated the effectiveness of using character substrings in an English NER.

The effectiveness of using dictionaries or gazetteers is shown in previous works. Florian et al. (2003) used location, person, and organization gazetteers in their NER framework and reported an error reduction in an extraction performance. Cohen and Sarawagi (2004) exploited a state, a city, a person, and a company dictionaries to improve NER. Jonnalagadda et al. (2013) used various medical resources in their NER system and showed an increase in an extraction performance of medical concepts. Automatic constructions of a dictionary/gazetteer are also examined. Kazama and Torisawa (2007) and Toral and Muñoz (2006) exploited Wikipedia to construct a dictionary/gazetteer that is useful for NER.

5 Discussion

5.1 Effects of Knowledge Resources

The use of terminology resulted to high precision recognizers. The best result in precision of 89.40% was obatained by only using the MEDIS terminologies, but its recall was the only one that did not show the statistically significant improvement against the baseline. The use of MedDRA terminology was similar to the use of MEDIS terminologies with a slightly higher recall and a slightly lower precision. Regardless of this similarity, terms that are present in NTCIR-10 MedNLP data are somewhat different between the two kinds of terminologies (Table 4). The percentages of terms that are not unique are about 40.4% and 45.7% for MedDRA/J and MEDIS BM & SSM respectively. We assume that even though the two kinds of terminologies are rather different in term presence, both kinds included similar information that is essential for NER.

The introduction of the external named entities (DSNE) resulted to a different result in certain extent compared to the terminology utilizations. The recall marked the second highest score of 84.18% but the precision was lower, although not statistically significant, than the baseline. We assume that symptom named entities and disease name entities in DS Corpus can be a clue to recognize complaints and diagnoses (high recall) but differences between them degraded the certainty of recognition (low precision).

5.2 Generality of Knowledge Resource Incorporation

The approach we took for the incorporation of terminology has a high generality. The approach requires only entries of a terminology. More rich contents like glosses or synonyms are not required. This characteristic makes the incorporation applicable to almost any kind of terminology.

The technique to introduce external named entities also has a high generality. The technique encodes named entity results as binary features for each entity type. This encoding can be done to almost any type of named entity. However, as mentioned in Section 5.1, the introduction of external named entity showed the defect in precision. This may be undesirable in some practical uses.

6 Conclusion

We presented a method that utilizes external medical knowledge into a state-of-the-art named entity recognizer. An evaluation using NTCIR-10 MedNLP data showed that the introduction of the medical knowledge resources improves the complaints and diagnoses extraction performance by about 2.00 in F_1 score. The best F_1 score 86.53 obtained in our method is comparable to top scoring results in Complaint and Diagnosis subtask of NTCIR-10 MedNLP.

The presented knowledge resource incorporation method has high generality, and its application is not restricted to the resources described in this paper. For example, a drug terminology can be incorporated to a medication extraction. This high generality suggests the promising future of a natural language processing in medical fields, where numerous knowledge resources are available.

References

Eiji Aramaki, Yasuhide Miura, Masatsugu Tonoike, Tomoko Ohkuma, Hiroshi Mashuichi, and Kazuhiko Ohe. 2009. TEXT2TABLE: Medical text summarization system based on named entity recognition and modality identification. In *Proceedings of the BioNLP 2009 Workshop*, pages 185–192.

Masayuki Asahara and Yuji Matsumoto. 2003. Japanese named entity extraction with redundant morphological analysis. In *Proceedings of HLT-NAACL 2003*, pages 8–15.

Elliot G. Brown, Louise Wood, and Sue Wood. 1999. The medical dictionary for regulatory activities (MedDRA). *Drug Safety*, 20(2):109–117.

William W. Cohen and Sunita Sarawagi. 2004. Exploiting dictionaries in named entity extraction: Combining semi-Markov extraction processes and data integration methods. In *Proceedings of KDD 2004*, pages 89–98.

Radu Florian, Abe Ittycheriah, Hongyan Jing, and Tong Zhang. 2003. Named entity recognition through classifier combination. In *Proceedings of CoNLL 2003*, pages 168–171.

Hiroto Imachi, Mizuki Morita, and Eiji Aramaki. 2013. NTCIR-10 MedNLP task baseline system. In *Proceedings of NTCIR-10*, pages 710–712.

Siddhartha Jonnalagadda, Trevor Cohen, Stephen Wu, Hongfang Liu, and Graciela Gonzalez. 2013. Evaluating the use of empirically constructed lexical resources for named entity recognition. In *Proceedings of CSCT 2013*, pages 23–33.

Junichi Kazama and Kentaro Torisawa. 2007. Exploiting Wikipedia as external knowledge for named entity recognition. In *Proceedings of EMNLP-CoNLL 2007*, pages 698–707.

Dan Klein, Joseph Smarr, Huy Nguyen, and Christopher D. Manning. 2003. Named entity recognition with character-level models. In *Proceedings of CoNLL-2003*, pages 180–183.

John Lafferty, Andrew McCallum, and Fernando Pereira. 2001. Conditional random fields: Probabilistic models for segmenting and labeling sequence data. In *Proceedings of ICML 2001*, pages 282–289.

Medical Information System Development Center, editor. 2012. *Hyojun Byomei Handobukku 2012 [Standard Disease Name Handbook 2012] (In Japanese)*. Shakai Hoken Kenkyujo, Inc.

Mizuki Morita, Yoshinobu Kano, Tomoko Ohkuma, Mai Miyabe, and Eiji Aramaki. 2013. Overview of the NTCIR-10 MedNLP task. In *Proceedings of NTCIR-10*, pages 696–701.

Eric W. Noreen. 1989. *Computer-Intensive Methods for Testing Hypotheses: An Introduction*. John Wiely and Sons, Inc.

Lev Ratinov and Dan Roth. 2009. Design challenges and misconceptions in named entity recognition. In *Proceedings of CoNLL-2009*, pages 147–155.

Antonio Toral and Rafael Muñoz. 2006. A proposal to automatically build and maintain gazetteers for named entity recognition by using Wikipedia. In *Proceedings of the Workshop on NEW TEXT Wikis and blogs and other dynamic text sources*, pages 56–61.

Clinical Vocabulary and Clinical Finding Concepts in Medical Literature

Takashi Okumura
National Institute of Public Health
taka@niph.go.jp

Eiji Aramaki
Kyoto University
eiji.aramaki@gmail.com

Yuka Tateisi
National Institute of Informatics
yucca@nii.ac.jp

Abstract

Clinical decision support systems necessitate a disease knowledge base, which comprises a set of clinical findings for each disease. To efficiently represent the findings, this paper explores the relationship between clinical vocabulary and findings in medical literature through quantitative and qualitative analysis of representative disease databases. Although the data volume and the analyzed features are limited, the observations suggested the following. First, there are sets of clinical findings that are essential for physicians, but the majority of findings in medical literature are not the essential ones. Second, deviation of term frequency for clinical findings vocabulary is minimal, and clinical findings require appropriate grammar for efficient representation of findings. Third, appropriate mapping of clinical findings with clinical vocabulary would allow the efficient expression of clinical findings.

1 Introduction

Clinical decision support systems necessitate a knowledge base of diseases, which comprises efficient representations of signs and symptoms for certain diseases. Such a knowledge base may efficiently represent a disease with relation to a set of predefined findings, such as *headache* and *nausea*. However, it is commonly observed that the derivatives of such findings become a diagnostic clue in the actual diagnosis process. For example, *morning headache* may suggest a tumor in the cranium, whereas cerebral hemorrhage may accompany *sudden headache*. These cases illustrate that, in clinical medicine, signs and symptoms modified with other elements may form a meaningful cluster that carries clinically valuable information.

In order to represent the "concepts of clinical findings" in an efficient manner, we are required to maintain appropriate vocabulary, as well as a variety of modifiers, such as *where*, *when*, and *how* the signs appear. For an ontology of diseases, the analysis of the relationship between such vocabulary for clinical findings and concepts of clinical findings is indispensable for efficient knowledge representation.

Accordingly, the paper performs quantitative and qualitative analysis of the vocabulary and the concepts of clinical findings in a couple of representative disease databases, OMIM (Online Mendelian Inheritance in Man) (McKusick, 2007) and Orphanet (Aymé and Schmidtke, 2007). In Section 2, we analyze the vocabulary of clinical findings, by assessing the impact of the vocabulary size against the coverage of words in descriptions of diseases. In Section 3, variations of clinical findings concepts are analyzed by assessing the expressions of clinical findings in the same texts. These analyses are followed by Section 4, which discusses the observations of the preceding sections. Section 5 summarizes a survey of related work, and Section 6 concludes the paper.

2 Variation of clinical findings vocabulary

In this section, the distribution of the terms for clinical findings is measured by taking simple statistics of terms used in OMIM and Orphanet. OMIM contains descriptions of approximately two thousand diseases in free format texts, and Orphanet has six thousand entries for diseases, including rare diseases. In the processing, MetaMap (Aronson and Lang, 2010; Aronson, 2001) is first applied to the texts, to extract the terms related to clinical findings. MetaMap is a tool to map phrases in a given medical literature text with UMLS (Unified Medical Language System) terminology (Lindberg et al., 1993), coupled with

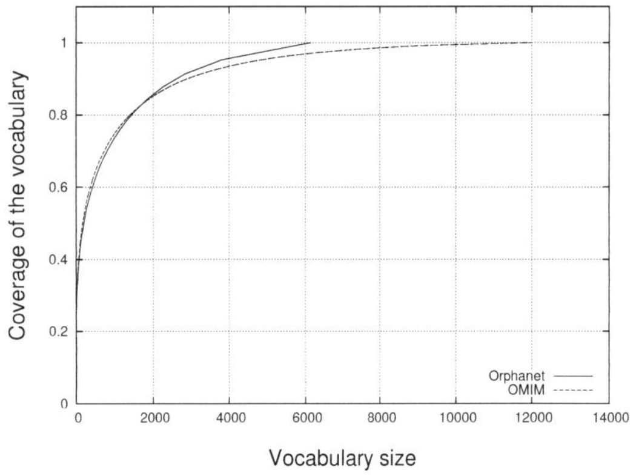

Figure 1: Vocabulary size and word coverage

Category	#	Ratio	Cumulative
Noun	254	18.6%	18.6%
Phrase			
Concept	476	34.8%	53.4%
Set of concepts	583	42.6%	96.0%
Sentence	55	4.0%	100.0%
Total	1368	100.0%	

Table 1: Grammatical categories of clinical findings in OMIM 20 documents

their semantic category. Then, the following subject categories of UMLS terms are used to extract clinical findings in the text: Acquired Abnormality, Anatomical Abnormality, Cell or Molecular Dysfunction, Congenital Abnormality, Disease or Syndrome, Embryonic Structure, Injury or Poisoning, Mental or Behavioral Dysfunction, Neoplastic Process, Sign or Symptom, and Virus. Lastly, the frequency of the extracted terms is measured, and the coverage of the symptomatic terms in the documents is calculated by increasing the vocabulary size in order of the term frequency.

Figure 1 shows the cumulative distribution of the terms. As illustrated, the coverage of the terms increases by adding terms for clinical findings into the vocabulary, and the top 2000 words covers 85% of the terms for clinical findings in the databases. Beyond this point, the coverage becomes less responsive to the addition of terms, because they are infrequently used in the target documents. The figure suggests that the difference between OMIM and Orphanet for the observed tendency is minimal.

To assess the size constraint of the vocabulary, we measured the percentage of simple words in the description of diseases. A clinical finding can be a word, such as *fever*, a phrase, such as *periodic fever*, or a sentence. If the portion of word findings is limited among all expression types, the unlimited vocabulary size by itself cannot achieve the appropriate representation of clinical findings. To estimate the upper bound of the contribution by the unlimited vocabulary, we analyzed the findings described in randomly selected OMIM documents (Document IDs: 108450, 113450 118450, 123450, 140450, 176450, 181450, 200450, 203450, 214450, 218450, 233450 236450, 244450, 248450, 259450, 265450, 267450, 305450, and 311450). The 20 texts included 1368 clinical findings in total. A sample phrase and a sentence are excerpted below:

"with most patients dying within 10 years of onset"
(OMIM 203450: Alexander Disease)

"No females manifested any symptoms."
(OMIM 305450: Opitz-Kaveggia Syndrome)

Table 1 shows the breakdown of the finding categories in the selected texts. The "Noun" category is for single word. The "Phrase" category is for phrases, which comprise phrases that represent either a concept, or a set of concepts. "Concept" includes phrases that can be mapped to a clinical concept, such as "mental retardation" and "Tetralogy of Fallot". "Set of concepts" is for phrases that are mapped to multiple concepts. As the table illustrates, the noun category accounts for only 18.6% of the expressions for clinical findings. Even if appropriate terminologies cover simple concepts for clinical findings (34.8%), they share only 53.4% of the entire findings and 46.6% of the expressions still necessitate phrases and sentences. Although the distinction of a concept and a set of concepts can be ambiguous in some cases, this tendency suggests that even the unlimited vocabulary cannot appropriately express all the clinical findings because the portion for vocabulary is limited in the actual descriptions of diseases.

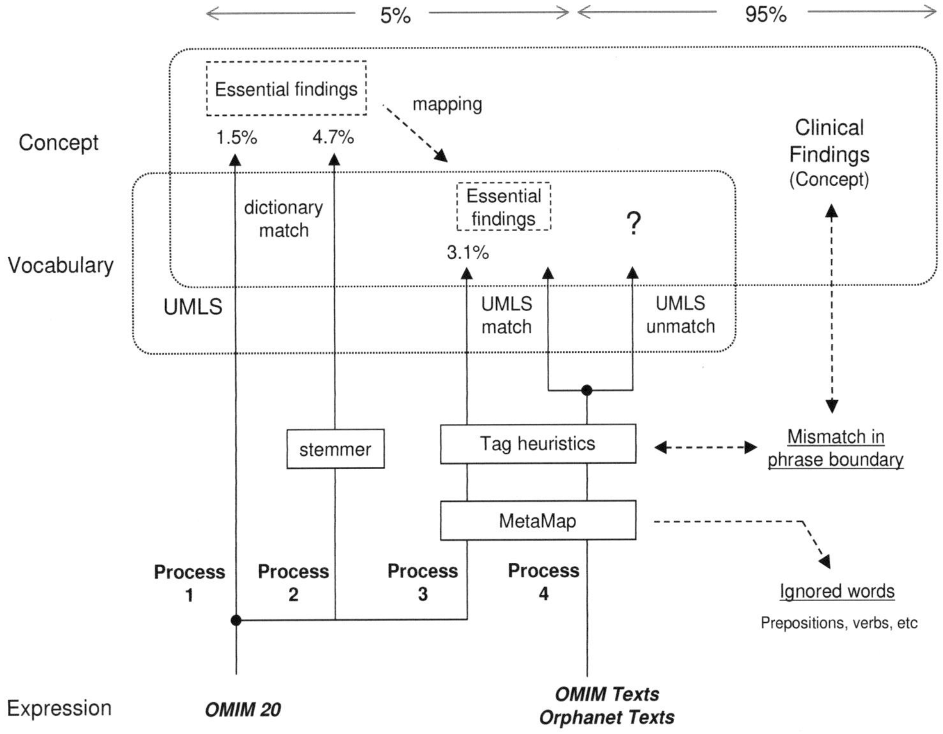

Figure 2: Coding of clinical findings concepts and the vocabulary

3 Variation of clinical findings concepts

The last section explored the vocabulary, required to express clinical findings. This section, inversely, examines clinical finding concepts in the descriptions of diseases. For this purpose, we used a set of clinical findings, compiled by Dr. Keijiro Torigoe for his rudimentary diagnostic reminder system (Torigoe et al., 2003). The dataset comprises 597 clinical manifestations, which include common signs and symptoms as well as entries for typical laboratory examinations results, such as high white blood cell counts and low platelets. The analysis in this section utilizes the essential findings for physicians to code clinical findings in the annotated OMIM texts. The entire setting is illustrated in Figure 2.

First, we performed dictionary matching of the essential finding, against the annotated OMIM texts (Process 1, Figure 2). The simple dictionary match showed that the essential findings accounted for only 1.5% of the annotated elements. Second, we applied a stemmer, Snowball (Porter, 2001), before the matching, which increased the recall to 4.7% (Process 2, Figure 2). Third, for further performance gain, we processed the essential findings with MetaMap and compiled the set of 586 findings in *UMLS Concept Unique Identifiers (CUI)*. Then, we performed the matching against the result of the MetaMap processing on the OMIM texts, which resulted in a 3.1% match (Process 3, Figure 2).

The three stages illustrated that the essential findings account for only 5% of the clinical findings in the sample documents, and the failure analysis suggested the following. First of all, MetaMap mostly ignores prepositions and verbs, which constitute essential parts in the expression of the clinical findings. Second, MetaMap segments the texts into minimum phrases that have corresponding CUI. However, annotators with a clinical background tend to group multiple phrases together, because they carry meaningful information as clinical findings. This results in the further mismatch between the MetaMap output and the concepts of clinical findings. Lastly, the dataset of essential findings could have missed some frequent terms.

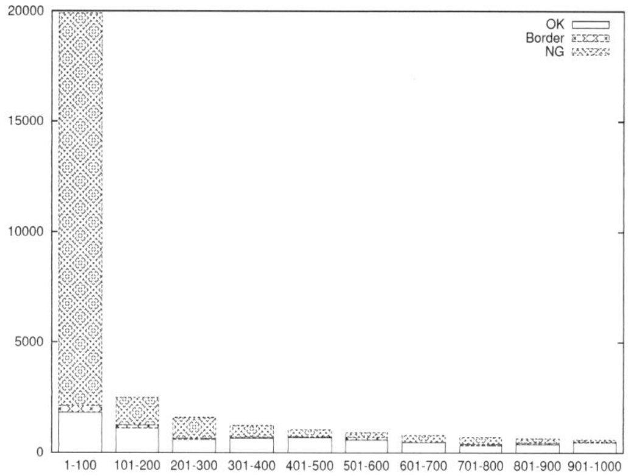

Figure 3: Word count of unmatched terms in Orphanet

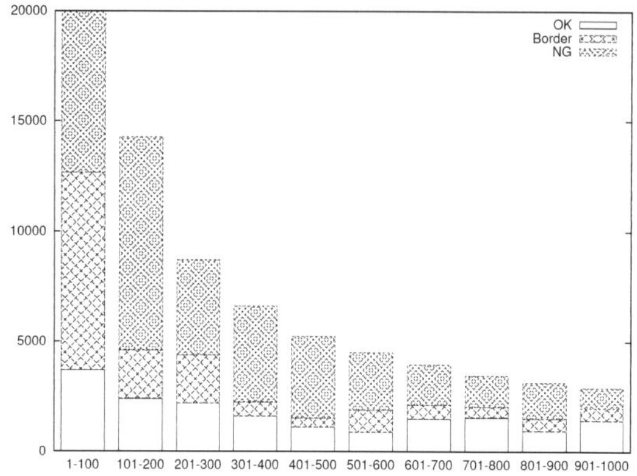

Figure 4: Word count of unmatched terms in OMIM

To verify the last conjecture and to detect unknown frequent findings, we analyzed the failure cases of UMLS matching. For this purpose, we applied the same processing on the entire OMIM and Orphanet texts and extracted unmatched cases (Process 4, Figure 2). The output was sorted in order of frequency and grouped into 10 clusters, each of which contained 100 words. Figures 3 and 4 show the word count of terms in each cluster (Note the range of y-axis is limited). As illustrated, the first clusters of both graphs exhibited striking peaks (Orphanet: 19898 counts and OMIM: 86085 counts) for the top 100 words. However, a detailed look revealed that half of the terms were useless (NG class), because they are functional terms that were mistakenly included by the tag heuristics. The class Border denotes borderline cases, which are clinical terms, but which do not carry clinical meaning in the context, such as *(the) disease* and *(the) symptom*.

Accordingly, we extracted the acceptable OK terms in the top 1000 unmatched words and measured their contribution to the frequency distribution of the terms in the target documents. As Figure 5 illustrates, the contribution of the unmatched words is limited: the top 100 unmatched words accounted for 6.8% of the findings word count for Orphanet documents, and 3.1% of the findings word count for OMIM documents. In all, the 1000 unmatched words contained 543 words for Orphanet and 278 words for OMIM, which accounted for 14.6% and 4.8% in the entire word count, respectively. The Orphanet cases outperformed the OMIM cases, which could be partially attributed to the difference in word counts (49,342

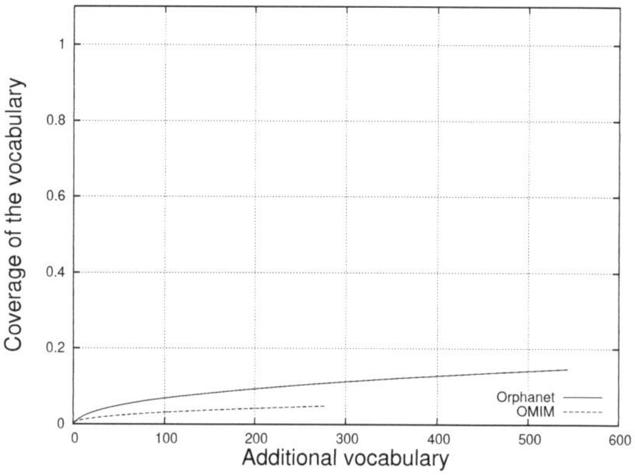

Figure 5: Additional vocabulary and contribution to the word coverage

against 361,566). The majority of the cases in the frequent unmatched words were technical terms, such as *microcephaly*, *hypertelorism*, and *facial dysmorphism*, which could be frequent for certain classes of genetic disorders but clinically uncommon.

The observation suggests that the number of essential clinical findings is approximately several hundreds, far below a thousand. These findings account for just a few percent of clinical findings documented in the descriptions of diseases in representative disease databases. The concepts of clinical findings in medical literature are diverse. Although some of the clinical finding concepts might be well-known, such as anatomical and congenital anomalies, most of them are clinically uncommon and do not appear often in literature. Although manual matching might increase the percentage, it is not likely that the overall picture would change.

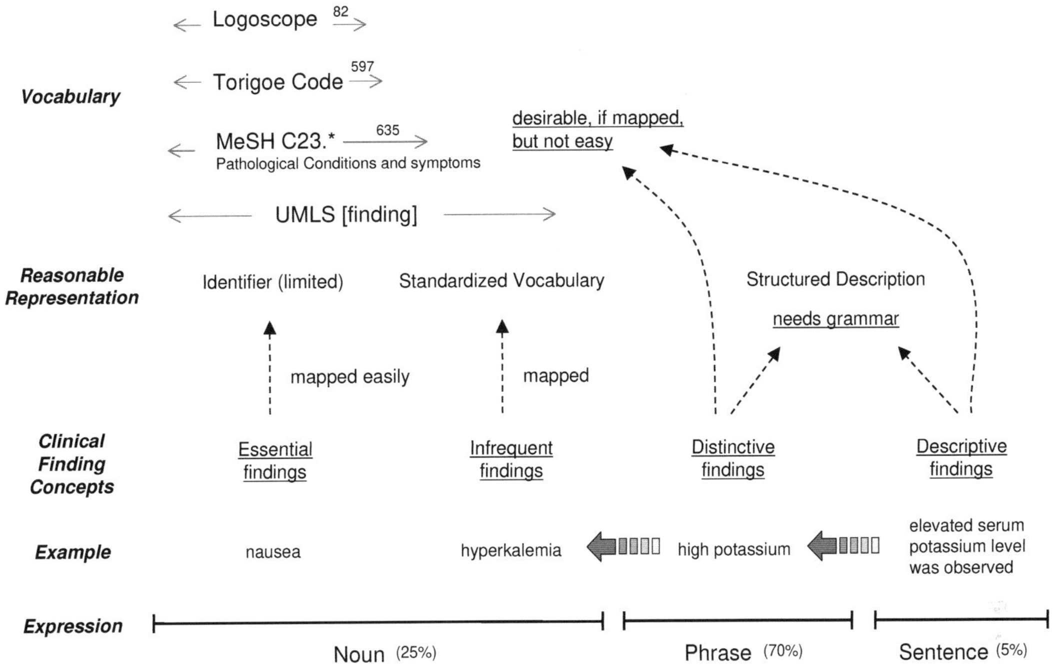

Figure 6: Clinical findings concepts and vocabularies

4 Discussion

This paper explores the relationship between clinical vocabulary and clinical findings through the analysis of disease descriptions compiled in OMIM and Orphanet. Figure 6 summarizes the observations. The essential findings, found as nouns in literature, are limited in number. This is supported by other datasets. For example, Logoscope (Nash, 1960) is a manually operated diagnostic tool called the "diagnostic slide rule", and it defines a set of 82 essential findings. MeSH (National Library of Medicine, 1963) has a category for essential findings named "pathological conditions and symptoms", which includes 635 terms. Infrequent findings are the other type of nouns, which can be mapped to corresponding terms in standardized vocabulary. There are other types of clinical findings, distinctive and descriptive findings, although the classification is tentative. Distinctive findings are those expressed by phrases, and descriptive findings are the most complex findings, which can be expressed only by sentences.

It is desirable that the findings are properly mapped onto the clinical findings vocabulary. However, it is not an easy task to map them, as suggested throughout the paper. For appropriate representation of clinical findings concepts, vocabulary alone cannot bridge the gap, and a grammar with appropriate descriptive power is indispensable. However, there is a tradeoff between the descriptive power and the cost for knowledge acquisition and representation.

mental retardation
intellectual deficiency
intellectual impairment
language delay
learning difficulties
poor school performance
delayed psychomotor development
psychomotor delay
psychomotor retardation

Table 2: Expressions for mental retardation

The knowledge acquisition of clinical findings is still a challenging task for Natural Language

Processing (NLP). Physicians may describe a finding in a sentence, which is common for pathological and radiological findings. Such a finding might have a corresponding term, and an elaborated system might cleverly map the sentence into a standardized vocabulary. However, this process involves various tasks, such as processing of negation, dependency, ambiguity, and abstraction, most of which are still unreliable for clinical use at this point. Even mapping of phrases is a challenge. For example, physicians may describe the concept "mental retardation" in diverse ways. Table 2 denotes how the concept is expressed in OMIM and Orphanet. Although MetaMap is a useful tool for mapping clinical terms, it still falls short of the required performance, to map sentences and phrases into standardized vocabulary.

The high cost of knowledge acquisition also applies to knowledge representation. Structured description of knowledge requires a grammar, which also burdens the data utilization process. Accordingly, it would be beneficial to reduce the cost for data representation, in addition to the improvement of knowledge acquisition performance. In this regard, physicians may express findings in phrases and sentences, when the findings are unfamiliar, or when they do not recall the appropriate terms even if one exists that corresponds to the concept. Examples include laboratory findings as illustrated at the bottom of Figure 6. Because a noun is the simplest form of knowledge, mapping of sentences and phrases into the terms might contribute to reducing the representation cost as well as the cost for data utilization.

5 Related works

Numerous research efforts have been made in the field of Natural Language Processing toward precise acquisition and representation of knowledge in clinical medicine.

First of all, there is a class of works aimed at boosting the accuracy of finding clinical manifestations in medical texts. Since the pioneering work (Sneiderman et al., 1996) in this problem domain, there have been various studies that investigated basic technologies required for accurate mining. Chapman proposed negation detection (Chapman et al., 2001) for clinical texts, which was extended to context handling methods (Chapman et al., 2007). Other groups focused on knowledge acquisition of diseases (Achour et al., 2001; Aleksovska-Stojkovska and Loskovska, 2010).

Second, researchers have worked on the knowledge representation issue for clinical findings. In addition to UMLS (Lindberg et al., 1993), which is used in this paper, SNOMED-CT (International Health Terminology Standard Development Organisation, 2001), MeSH (National Library of Medicine, 1963), OpenGALEN (Rector et al., 2003), and MedDRA (MedDRA Maintenance and Support Services Organization, 2007) have been used for the representation of clinical concepts. There are other studies in this domain (Sager et al., 1994; Cimino, 1991; Kong et al., 2008; Peleg and Tu, 2006).

Lastly, there have been several lines of work that explored the tools for information extraction on clinical reports. For example, (Friedman et al., 1994) developed Medical Language Extraction and Encoding (MedLEE) to encode clinical documents in a structured form. The Mayo Clinic also developed a similar NLP system (cTakes) (Savova et al., 2010) for clinical reports and TEXT2TABLE (Aramaki et al., 2009) targeted Japanese discharge summaries.

6 Conclusion

This paper investigated clinical vocabulary and clinical finding concepts. Because the analysis is made with a limited set of data, further study is required for more rigorous proof. Nevertheless, the current observations suggest the following.

First, there are essential findings for physicians and, in medical literature, the majority of the findings do not fall into the category. This observation is consistent with the fact that annotated findings tend to span multiple words.

Second, the deviation of the term frequency for clinical findings vocabulary is minimal, and the vocabulary alone cannot express all the clinical findings. Even with the UMLS terminology, the expressive power is limited, which necessitates an appropriate grammar for structured descriptions of findings.

Third, knowledge acquisition of clinical findings is costly, and the grammar would escalate the cost for representation, as well as the cost for data utilization. However, appropriate mapping of clinical findings and clinical vocabulary, particularly for infrequent terms, might contribute toward expressing clinical findings without increasing the cost for representation and for utilization.

References

Soumeya L Achour, Michel Dojat, Claire Rieux, Philippe Bierling, and Eric Lepage. 2001. A umls-based knowledge acquisition tool for rule-based clinical decision support system development. *Journal of the American Medical Informatics Association*, 8(4):351–360.

Liljana Aleksovska-Stojkovska and Suzana Loskovska. 2010. Clinical decision support systems: Medical knowledge acquisition and representation methods. In *2010 IEEE International Conference on Electro/Information Technology (EIT)*, pages 1–6. IEEE.

Eiji Aramaki, Yasuhide Miura, Masatsugu Tonoike, Tomoko Ohkuma, Hiroshi Mashuichi, and Kazuhiko Ohe. 2009. Text2table: Medical text summarization system based on named entity recognition and modality identification. In *the Human Language Technology conference and the North American chapter of the Association for Computational Linguistics (HLT-NAACL2009) Workshop on BioNLP*, pages 185–192.

Alan R Aronson and Francois-Michel Lang. 2010. An overview of MetaMap: historical perspective and recent advances. *Journal of the American Medical Informatics Association*, 17(3):229–36.

Alan R Aronson. 2001. Effective mapping of biomedical text to the UMLS Metathesaurus: the MetaMap program. In *AMIA Annual Symposium*, pages 17–21.

Ségolène Aymé and Jorg Schmidtke. 2007. Networking for rare diseases: a necessity for europe. *Bundesgesundheitsblatt-Gesundheitsforschung-Gesundheitsschutz*, 50(12):1477–1483.

Wendy W Chapman, Will Bridewell, Paul Hanbury, Gregory F Cooper, and Bruce G Buchanan. 2001. A simple algorithm for identifying negated findings and diseases in discharge summaries. *Journal of biomedical informatics*, 34(5):301–310.

Wendy W Chapman, John Dowling, and David Chu. 2007. Context: An algorithm for identifying contextual features from clinical text. In *Biological, translational, and clinical language processing (BioNLP2007)*, pages 81–88.

James J Cimino. 1991. Representation of clinical laboratory terminology in the unified medical language system. In *Proceedings of the Annual Symposium on Computer Application in Medical Care*, page 199. American Medical Informatics Association.

Carol Friedman, Philip O Alderson, John HM Austin, James J Cimino, and Stephen B Johnson. 1994. A general natural-language text processor for clinical radiology. *Journal of the American Medical Informatics Association*, 1(2):161–174.

International Health Terminology Standard Development Organisation. 2001. SNOMED CT. http://snomed.org/.

Guilan Kong, Dong-Ling Xu, and Jian-Bo Yang. 2008. Clinical decision support systems: a review on knowledge representation and inference under uncertainties. *International Journal of Computational Intelligence Systems*, 1(2):159–167.

Donald Lindberg, Betsy Humphreys, and Alexa McCray. 1993. The unified medical language system. *Methods of Information in Medicine*, 32(4):281–91.

Victor A. McKusick. 2007. Mendelian inheritance in man and its online version, omim. *American journal of human genetics*, 80(4):588–604, April.

MedDRA Maintenance and Support Services Organization. 2007. *Introductory Guide, MedDRA Version 10.1*. International Federation of Pharmaceutical Manufacturers and Associations.

Firmin A. Nash. 1960. Diagnostic reasoning and the logoscope. *Lancet*, 276:1442–1446, December.

National Library of Medicine. 1963. Medical Subject Headings. http://www.nlm.nih.gov/mesh/.

Mor Peleg and Samson Tu. 2006. Decision support, knowledge representation and management in medicine. *Yearb Med Inform*, 45:72–80.

Martin Porter. 2001. Snowball: A language for stemming algorithms.

Alan Rector, Jeremy Rogers, Pieter Zanstra, and Egbert van der Haring. 2003. Opengalen: open source medical terminology and tools. In *AMIA Annual Symposium Proceedings*, page 982.

Naomi Sager, Margaret Lyman, Christine Bucknall, Ngo Nhan, and Leo J Tick. 1994. Natural language processing and the representation of clinical data. *Journal of the American Medical Informatics Association*, 1(2):142–160.

Guergana K Savova, James J Masanz, Philip V Ogren, Jiaping Zheng, Sunghwan Sohn, Karin C Kipper-Schuler, and Christopher G Chute. 2010. Mayo clinical text analysis and knowledge extraction system (cTAKES): architecture, component evaluation and applications. *Journal of the American Medical Informatics Association*, 17(5):507–513.

Charles A Sneiderman, Thomas C Rindflesch, and Alan R Aronson. 1996. Finding the findings: identification of findings in medical literature using restricted natural language processing. In *AMIA Annual Fall Symposium*, pages 239–43. AMIA.

Keijirou Torigoe, Gen'ichi Kato, and Yoshio Ohta. 2003. Computer-aided diagnosis supporting tool. *Japan Medical Journal*, (4120):24–32. (in Japanese).

Developing ML-based Systems to Extract Medical Information from Japanese Medical History Summaries

Shohei Higashiyama **Kazuhiro Seki** **Kuniaki Uehara**
Graduate School of System Informatics, Kobe University
{higashiyama@ai.cs., seki@cs., uehara@}kobe-u.ac.jp

Abstract

With the increase of the number of medical records written in an electronic format, natural language processing techniques in the medical domain have become more and more important. For the purpose of the development and evaluation of machine learning-based systems to extract medical information, we recently participated in the NTCIR-10 MedNLP task. The task focused on Japanese medical records and aimed at evaluating different information extraction techniques on the common data set provided by the organizers. We implemented our baseline system based on structured perceptron and have developed its extensions. In this paper, we describe our systems and report on the evaluation of and the analysis on their performance.

1 Introduction

In recent years, medical records have been increasingly written in an electronic format, which leads to a growing need for natural language processing (NLP) techniques in the medical domain. Specifically, information extraction (IE) techniques, such as named entity recognition (NER), are crucial as they serve as the basis of more intellectual and/or application-oriented tasks, including information retrieval and question answering.

Given the background, the NTCIR-10 MedNLP task (Morita et al., 2013) was recently held as a shared task to foster the NLP research for medical texts, specifically targeting Japanese. The participants of the task were provided with an annotated corpus consisting of 50 fictional medical history summary reports. The intended task was a type of NER and required the participants to identify patients' personal and medical information from the reports.

For the MedNLP task, we took part in the de-identification subtask and the complaint and diagnosis subtask summarized shortly by adapting an NER model to the medical domain. The model is based on structured perceptron (Collins, 2002) and was previously developed for the biomedical domain (Higashiyama et al., to appear).

This paper reports on the results of the structured perceptron-based model for the MedNLP task and presents their analysis. Additionally, conditional random fields (CRFs) (Lafferty et al., 2001), a popular model adopted by many participants of the task, are applied for comparison.

2 NTCIR-10 MedNLP Task

2.1 Dataset

The MedNLP task organizers prepared medical history summary reports of fictional patients written by physicians. The medical records consist of 50 documents and include 3,365 sentences. Two thirds of them (2,244 sentences) and remaining one thirds (1,121 sentences) are respectively provided as the sample set and the test set.

The sample set is annotated with personal and medical information about patients. The personal information includes *age*, *person's name*, *sex*, *time*, *hospital name* and *location* [1]. The medical information indicates *complaint and diagnosis* with a modality attribute that is taken to have four values: *positive*, *negation*, *suspicion* and *family*. Suppose that there is a mention of a particular symptom about a patient. Then, the expression representing the symptom would be annotated with the attribute value of *positive* if the patient has the symptom, *negation* if the patient does not have the symptom, *suspicion* if the patient is suspected of the symptom and *family* if a member

[1] A half of these tags in fact rarely appear in the sample set. The numbers of *persons' name*, *sex* and *location* tags are less than five while the numbers of remaining tags are respectively more than 50.

of the patient's family has a history of the symptom.

2.2 Task Description and Formulation

The NTCIR-10 MedNLP task mainly consisted of the following two subtasks.

1. De-identification (DI) task: identifying personal information about patients, such as ages and hospital names.

2. Complaint and diagnosis (CD) task: extracting patients' complaint and diagnosis by physicians and determining their modality status for the patients.

The performance of participants' systems for both subtasks was measured by the F-measure ($\beta = 1$), which is the harmonic mean of precision and recall.

These subtasks can be seen as NER tasks recognizing named entities and classifying them into predefined semantic classes. Named entities indicate particular expressions to be extracted, which are represented by proper nouns and technical terms. As for the DI task, this subtask can be formulated as classifying each word in a sentence into one of the labels consisting of a semantic class (e.g. *age*) and a chunk IOB tag, where I, O, and B respectively denote the inside, outside, and beginning of an entity. For example, if a word "64" in "64 years old" is assigned with a label "`B-age`", it means that the "64" is recognized as the beginning of an entity with a semantic class *age*. The CD task can be formulated likewise by regarding a *complaint and diagnosis* tag with a modality attribute *x* as a class *c-x*.

3 Description of Baseline System

For the MedNLP task, we applied structured perceptron (Collins, 2002), which is an online algorithm. Despite its simplicity, structured perceptron is reported to have performance that closely approximates that of support vector machines (SVMs), which has been applied successfully to various classification problems. In addition, we introduced a cost function into the perceptron framework to achieve higher performance, and used the model as our baseline system. The cost function is a type of cost-sensitive learning method which lowers the expected cost of misclassification.

In the following two sections, we describe the learning and prediction algorithms on an ordinary and a cost-sensitive version of structured perceptron.

3.1 Structured Perceptron

Let $\mathcal{X}$ be a set of instances and let $\mathcal{Y}_{\boldsymbol{x}}$ be a set of possible label sequences for an instance $\boldsymbol{x} \in \mathcal{X}$, where $\boldsymbol{x}$ denotes a token sequence (i.e., sentence) in the training or test data. Additionally, $\boldsymbol{y} \in \mathcal{Y}_{\boldsymbol{x}}$ denotes a possible label sequence of $\boldsymbol{x}$. $\mathcal{Y}_{\boldsymbol{x}}$ is equivalent to the direct product $\mathcal{L}^n$, where n is the length of $\boldsymbol{x}$ and $\mathcal{L}$ is a set of labels that includes labels such as `B-age` and `O`.

Learning on structured perceptron can be regarded as finding the weight vector $\boldsymbol{w} \in \mathbb{R}^d$ so that the discriminative function f predicts the correct label sequences of instances. The discriminative function $f : \mathcal{X} \to \mathcal{Y}$ is defined as

$$f(\boldsymbol{x}, \boldsymbol{y}) = \langle \boldsymbol{w}, \Phi(\boldsymbol{x}, \boldsymbol{y}) \rangle \,,$$

where $\langle \cdot, \cdot \rangle$ denotes an inner product of two arguments and $\Phi(\boldsymbol{x}, \boldsymbol{y}) \in \mathbb{R}^d$ is the feature vector of $\boldsymbol{x}$ and $\boldsymbol{y}$.

The prediction $\hat{\boldsymbol{y}}$ for $\boldsymbol{x}$ is the output of f as in

$$\hat{\boldsymbol{y}} = \operatorname*{argmax}_{\boldsymbol{y} \in \mathcal{Y}} f(\boldsymbol{x}, \boldsymbol{y}) \,. \qquad (1)$$

During learning on the training data, we receive a training instance $\boldsymbol{x}_t$ on each round t, and output its prediction $\hat{\boldsymbol{y}}_t$ by Eq. (1). Then, $\boldsymbol{w}$ is updated by Eq. (2) if the prediction $\hat{\boldsymbol{y}}_t$ differs from the correct label sequence $\boldsymbol{y}_t$:

$$\boldsymbol{w}^{t+1} \leftarrow \boldsymbol{w}^t + \Phi(\boldsymbol{x}_t, \boldsymbol{y}_t) - \Phi(\boldsymbol{x}_t, \hat{\boldsymbol{y}}_t) \,, \qquad (2)$$

where $\boldsymbol{w}^t$ is the weight vector on round t. Learning is iterated through all the training instances T times. Label sequences of test instances can be predicted by Eq. (1) in the same manner as training instances.

3.2 Cost-Sensitive Structured Perceptron

In addition to use of structured perceptron, we exploited information on distance between a correct and a candidate label sequence of each training instance during learning based on cost-sensitive learning of an ML framework for lowering misclassification cost. Cost-sensitive approaches were, for example, applied to semantic role labeling on the study by Johansson and Nugues (2008), which used passive-aggressive (Crammer et al., 2006), and to part-of-speech tagging on that by Song et al. (2012), which used multiclass SVMs.

The cost-sensitive learning algorithm on structured perceptron updates the weight vector $\boldsymbol{w}$ using $\tilde{\boldsymbol{y}}_t$ defined below instead of $\hat{\boldsymbol{y}}_t$ in Eq. (2).

$$\tilde{\boldsymbol{y}}_t = \underset{\boldsymbol{y}\in\mathcal{Y}}{\text{argmax}}\ f(\boldsymbol{x}_t, \boldsymbol{y}) + \alpha\rho(\boldsymbol{y}_t, \boldsymbol{y}) \quad (3)$$

In Eq. (3), $\rho : \mathcal{Y}\times\mathcal{Y} \to \mathbb{N}\cup\{0\}$ is the cost function which returns a larger value for larger distance between $\boldsymbol{y}_t$ and $\boldsymbol{y}$, and α is a parameter that is taken to have a positive real number. Here, we define the cost function ρ as

$$\rho(\boldsymbol{y}_1, \boldsymbol{y}_2) = \sum_{i=1}^{|\boldsymbol{y}_1|} \delta(y_1^{(i)}, y_2^{(i)}) ,$$

where $|\boldsymbol{y}|$ denotes the length of the vector $\boldsymbol{y}$ and the function $\delta : \mathcal{L} \to \{0, 1\}$ is defined as

$$\delta(y_1, y_2) = \begin{cases} 0 & (y_1 = y_2) \\ 1 & (y_1 \neq y_2) \end{cases} .$$

In the cost-sensitive learning framework, the weight vector can be updated to the reserve margin $\alpha\rho(\boldsymbol{y}_t, \tilde{\boldsymbol{y}}_t)$ using $\tilde{\boldsymbol{y}}_t$ instead of $\hat{\boldsymbol{y}}_t$. That is,

$$\boldsymbol{w}^{t+1} \leftarrow \boldsymbol{w}^t + \Phi(\boldsymbol{x}_t, \boldsymbol{y}_t) - \Phi(\boldsymbol{x}_t, \tilde{\boldsymbol{y}}_t) .$$

3.3 Features

The following features were used in the experiments for both subtasks:

- tokens in the window of size two around the current token and
- the part-of-speech (POS) tag, the subtype of POS tag, the lemma and the pronunciation of the current token.

We applied the Japanese morphological analyzer MeCab (Kudo et al., 2004) (version 0.996) with the IPA dictionary [2] (version 2.7.0) to word segmentation and used the output of MeCab for each sentence as the latter features.

4 Evaluation and Discussion

4.1 Evaluation of Baseline System

Parameter Setting

We determined the optimal value of parameter α in Eq. (3) and the number of iterations T using the sample set as follows.

1. We used 90% of the sample set as the learning set and the remaining 10% as the validation set.

2. Varying the value of α and increasing the value of T, we learned a model for particular α and T on the learning set and evaluated it on the validation set.

3. Values of α and T that yielded the best F-measure were regarded as optimal.

Consequently, the optimal α and the number of iterations T were respectively set to 30 and 20. By use of the cost function, both precision and recall on the validation set improved by around four points, compared with the method without the function. We used these values for producing our official runs on the test set submitted to the MedNLP organizers.

Results on Test Set

Table 1 shows the performance of our system using the test set. Table 1 (a) shows the overall performance and Table 1 (b) shows the performance of each entity class. The performance was measured by precision, recall, the F-measure ($\beta = 1$), and accuracy. Recall was always lower than precision for all classes of both tasks, and especially lower in the family and the suspicion classes, which led to degraded F-scores. In addition, the lower performance for the total on the CD task than 2-way indicate difficulty of modality classification.

4.2 Error Analysis of Baseline System

For error analysis, we evaluated our system on the sample set using a five-fold cross-validation method. Subsequently, we analyzed the results on the validation sets for five iterations. As compared with the performance on the test set, the performance on the validation sets was worse by several points for the CD task, and almost equivalent for the DI task. The reason of the former is the fewer training instances, and that of the latter was that the targeted entities for the DI task have much in common as we discuss shortly.

Analysis on De-identification Task

Despite the smaller number of positive instances of entity classes for the DI task than that for the CD task, the performance for the former classes was relatively high on the whole. The

[2] http://chasen.naist.jp/stable/ipadic/

Table 1: Results of both de-identification (DI) task and complaint and diagnosis (CD) task on the test set. The "2-way" is a result of recognition of complaint/diagnosis or not. The "total" is a result including classification of modality classes. P, R, F and A indicate precision, recall, F-measure ($\beta = 1$), and accuracy, respectively.

(a) Overall performance on test set.

subtask	P	R	F	A
DI	82.09	76.39	79.14	99.38
CD (2-way)	82.37	72.29	77.00	95.48
CD (total)	74.72	65.58	69.86	94.50

(b) Performance of each entity class on test set.

subtask	class	P	R	F
DI	age	80.65	78.12	79.37
	time	84.56	81.56	83.03
	hospital	72.73	63.16	67.61
CD	c-positive	72.87	67.04	69.83
	c-negation	82.35	68.02	74.50
	c-suspicion	55.00	36.67	44.00
	c-family	66.67	36.36	47.06

reason is that a large portion of these entities fit typical patterns. For example, over 70 percents of the instances of the age class in the sample set match a simple regular expression, "[１-９]?[０-９]歳[時頃(ごろ)]?[ー〜(から)(より)(まで)]?" ("[(from)(to)]?(about)?[1-9]?[0-9](years old)"). For misclassified cases, we found two major types of errors across all classes in this task: (1) recognition of incorrect boundaries of entities; and (2) undetection of entities (false negatives).

Specifically, the most frequent errors on the age class was found to be the first type, such as "４７歳" (47 years old) for a correct boundary "２７歳〜４７歳" (27 to 47 years old) and "１０代' ' (10s) for "１０代前半" (early 10s). Because words or expressions co-occurring with or including ages themselves as numerical values are limited, it may be effective to fix system outputs by rule-based post-processing.

On the other hand, most errors on the hospital class was the second type. For example, entities such as "同院" (the hospital) and "総合病院" (general hospital) were often undetected. The reason is that these words rarely appeared in the sample set in contrast to frequently appearing words, such as "当院" (our hospital) and "近医" (local hospital), which were correctly detected.

As for the time class, both types of errors were often observed. A large portion of boundary errors were recognizing narrower scopes for entities than their correct ones, e.g., "１０月２９日" (October 29) against a correct boundary "１０月２９日夕刻まで" (until the evening on October 29). Many false negatives were found to be expressions using slashes, such as "７／２０". More formal expressions, such as "７月２０日" (July 20), are more often used in the sample set. For dealing with the errors of the hospital and the former type of the time, constructing and using dictionaries composed of expressions which often constitute or co-occur with those type of entities may be beneficial. For the latter type of the time, rule-based post-processing may be effective, similarly to the age class.

Analysis on Complaint and Diagnosis Task

In addition to the two types of errors discussed for the previous task, there were mainly two types of errors in detecting complaint entities: (3) misclassification of the modality classes; and (4) misdetection of non-entities (false positive).

The most frequent errors were undetection of entities through all classes, and this type of errors frequently observed in the positive and the negation classes. In order to reduce such false negatives and improve recall, we plan to use external knowledge resources such as public dictionaries in future work.

The second most frequent errors were misclassification of entities whose boundaries were correctly recognized. They accounted for a major portion of errors on the three classes except the positive class. Especially, the low performance on the family and the suspicion classes was due to misclassification in addition to undetection which occur similarly as the other modality classes. For these modality classes, it was found that there exist typical keywords which often co-occur with entities. Entities of the family class co-occur with family relation names. In particular, most of them in the sample set co-occur in itemized sentences, such as "父：心筋梗塞" (Father: cardiac infarction). Entities of the negative class and the suspicion class occur ahead of expressions of negations, such as "なし" (be absent), and expressions of uncertainty, such as "考えられる" (be concerned), "疑いがある" (be suspected), and "可能性がある"

Input:	薬剤性肺炎の可能性を考え
(consider the possibility of drug-induced pneumonia)	
⇓	
Output:	薬剤/性/肺炎/の/可能/性/を/考え

Figure 1: An example of a parsed sentence including a suspicion entity by MeCab. The underlined part in the input sentence indicates an entity annotated with the suspicion class. The parts segmented by slashes in the output indicate words segmented by the tagger.

(be possible).

However, our system could not exploit these keywords because of the limited window size of two around the current token, and entities often occur at a distance from keywords, especially in the suspicion class. For example, Figure 1 shows an input sentence containing a suspicion entity "薬剤性肺炎" (drug-induced pneumonia) and its parsed output by the MeCab morphological analyzer. Two out of three tokens constituting the entity (i.e., "薬剤" (drug) and "性" (-induced)) are more than two tokens away from the uncertainty keywords (i.e., "可能", "性" (possibility) and "考え" (concern)). To improve classification performance for modality classes, specifically recall, it is crucial to increase the window size to, for example, sentence boundaries. Alternatively, it may be effective to take advantage of dependency parsing.

The other causes of the observed errors were incorrect boundary errors and misdetection errors. The reasons require a further study.

4.3 Post-submission Experiments

To achieve higher performance, we have developed our medical information extracting systems also after implemented and submitted our baseline system. Specifically, we used CRFs as an alternative ML algorithm to structured perceptron. Moreover, we introduced domain-specific terms in medical fields into the default dictionary of the morphological analyzer.

In the following subsections, we describe the above conversion and extension from the baseline system and the experiments on those.

Alternative ML Algorithm: Conditional Random Fields

To improve the performance of the baseline system, we employed CRFs (Lafferty et al., 2001) as an alternative ML algorithm. CRFs are extensions of maximum entropy to structured prediction. Additionally, the algorithm has been widely applied to both NER (McCallum and Wei, 2003; Settles, 2004; Finkel et al., 2005) and other NLP tasks, such as part-of-speech tagging (Lafferty et al., 2001), noun phrase chunking (Sha and Pereira, 2003) and morphological analysis (Kudo et al., 2004). Particularly, we utilized CRF++ [3], which is an open source implementation of CRFs and allows easy customizability of features by describing in the feature template file. We used the same features as those in the baseline system.

Use of Medical Lexicon

When analyzing texts in a specific domain, morphological taggers with default dictionary in general domain often unsuccessfully analyze sentences that contain domain-specific terms. Consequently, they make errors attributed to unknown words in word segmentation or other processing such as POS tagging and pronunciation prediction. These errors can be negatively affect on NER that is a higher-level task than morphological analysis. Then, we enhanced the regulation dictionary of MeCab by addition of domain-specific terminology from life science dictionary (LSD) (Kaneko et al., 2003), which consists of a broad range of life science terms such as names of anatomical concepts, biological organisms, diseases and symptoms.

By addition of a domain-specific dictionary, not only the morphological tagger can achieve tagging error reduction, but also finely segmented morphemes that are component of domain-specific terms tend to be segmented more coarsely because expressions contained in the dictionary are more frequently regarded as one morpheme. For instance, "薬剤性肺炎" (drug-induced pneumonia) is segmented into "薬剤" (drug), "性" (-induced) and "肺炎" (pneumonia) before the addition of terms in LSD to the original dictionary and into "薬剤性肺炎" after the addition. Similarly, "ＰＩＰ関節裂隙狭小化" (joint space narrowing at the proximal interphalangeal (PIP) joints) is seg-

[3] `http://crfpp.googlecode.com/svn/trunk/doc/index.html`

Table 2: Comparison of systems based on two algorithms with or without the enhanced dictionary using the sample set. SP denotes cost-sensitive structured perceptron and dic indicates using the enhanced dictionary.

(a) Performance for de-identification (DI) task.

system	P	R	F
SP	82.72	86.97	84.79
SP+dic	84.06	86.02	85.03
CRFs	91.01	82.32	86.45
CRFs+dic	89.26	82.61	85.81

(b) Performance for complaint and diagnosis (CD) task.

system	P	R	F
SP	66.29	72.76	69.37
SP+dic	65.05	77.02	70.53
CRFs	78.85	68.26	73.17
CRFs+dic	81.91	66.06	73.14

mented into "ＰＩＰ", "関節" (joints), "裂隙" (space), "狭小" (narrow) and "化" (-ing) before and into "ＰＩＰ関節", "裂隙" and "狭小化" after. The latter segmentation can be beneficial for exploiting information about strings distant from the token in question in the case of fixed window size around the token. Therefore, in addition to reduction errors in morphological analysis, NER systems can obtain benefit from coarse segmentation, by use of the tagger with the richer language resource.

Results and Discussion

To measure the performance of CRFs, which we used as an alternative algorithm to structured perceptron, and to evaluate the effectiveness of the enhanced dictionary, we compared four systems based on the two algorithms with or without the enhanced dictionary. Table 2 shows the results on the sample set using five-fold cross-validation. Table 2 (a) and (b) show the overall performance for the DI task and the CD task, respectively. For both subtasks, while recall of structured perceptron was higher than that of CRFs, CRFs outperformed structured perceptron by around 10 points in terms of precision. Additionally, CRFs also outperformed by a few points in terms of F-measure.

The both algorithms consider the overall sequence of tokens when predicting their labels, but they defer in the respective training methods. More precisely, structured perceptron minimizes the loss defined by the difference between correct and predicted label sequences. This process can be regarded as the training by a simple (sub) gradient method with fixed step size, which is a first-order gradient method. On the other hand, CRFs are trained by maximizing the log-likelihood of a given training set. The implementation of CRFs used in our experiment was based on limited-memory BFGS (L-BFGS), which is a second-order gradient method. We believe that the more sophisticated optimization algorithm of CRFs resulted in the higher performance. In fact, Sha and Pereira (2003) empirically showed that CRFs based on second-order methods, such as L-BFGS and conjugate gradient, outperformed structured perceptron on a noun phrase chunking task.

Contrary to our expectation, use of the morphological analyzer with enhanced dictionary had a little or negative effect for the performance of both algorithm and for both subtasks, except that recall of structured perceptron for the CD task was improved. We believe that this result was due to loss of common characteristics among segmented tokens. Focusing on the complaint entity "薬剤性肺炎" (drug-induced pneumonia), various expressions occur in the sample set preceding "肺炎" (pneumonia), e.g. "細菌性" (bacterial), "間質性" (interstitial), "器質化" (organizing), "強膜炎" (pleuritic) and "ニューモシスチス" (Pneumocystis), in addition to "薬剤性" (drug-induced). Furthermore, there are variety of entities containing expressions that co-occur with "肺炎", e.g. "薬剤性肺障害" (drug-induced pulmonary disorder), "細菌感染" (bacteria infection), "器質化血栓" (organizing thrombus), "胸膜炎" (pleuritis) and "ニューモシスチス・カリニ" (Pneumocystis carinii). As we discussed previously, morphemes tend to be segmented more coarsely after augmented terms in the dictionary of the morphological analyzer. Then, entities enumerated above became to be recognized as distinct tokens without common characteristics, by segmented to one or a little larger numbers of morphemes. We consider that this affected the performance negatively and disturbed learning of classifiers.

To fix this problem, it may be effective to use prefix and suffix features derived form expressions that are often contained by or co-occurred with entities. After the processing, classifiers may come to be able to exploit information about strings that are distant from the current token and to obtain benefit by reduction errors in morphological analysis.

5 Related Work

To the NTCIR-10 MedNLP task, both rule-based and ML-based approaches were applied among the participants. Almost all systems for the DI task and over a half of all systems for the CD task were based on ML, especially supervised learning. It should be note that greater part of systems that achieved higher performance were based on ML and moreover a large portion of them employed CRFs. Specifically, systems of the top three teams for the CD task and of the second and third ranked teams for the DI task were based on CRFs. By contrast, the system that had the highest performance for the DI task was a rule-based approach. As other ML-based approaches than CRFs, structured perceptron, language models and bootstrapping were applied.

As to features, general-purpose NER features were widely applied, such as word surface (token) and POS features. Pronunciation and character type features were also used. Besides, domain-specific features including dictionary matching features or heuristic features of data-specific expressions were used. These features are derived from medical knowledge resources such as LSD and MEDIS standard masters [4], or manually constructed lexica consisting of expressions that are specific to each entity class. Among the features incorporated in the ML-based systems, particularly, those that achieved higher performance, dictionary or heuristic features provided high benefit for their performance. Specifically, Laquerre et al. (Laquerre and Malon, 2013) reported that heuristic features for the DI task improved the F-measure by around three points and heuristic and dictionary features for the CD task improved by around 4.5 points. Miura et al. (Miura et al., 2013) also reported that dictionary features for the CD task improved the F-measure by around two points.

Nevertheless the limited size of the dataset, the overall performance for the subtasks of the top systems were high: they achieved over 90% and 75% F-measure for the DI task and the CD task, respectively. As regards the performance for each entity type, that for the family entities were over 80% F-measure, which is highest of all entity types for the CD task, in spite of smaller numbers of entities in the sample set. This is due to the features for the family class such as family names could capture the characteristics of this entities well. By contrast, the F-measure was only around 50% for the suspicion entities, which occurred less frequently similarly to the family entities. This suggests that the suspicious expressions used for extracting the suspicion entities (e.g. "疑い" (suspicious) and "可能性" (possibility)) were insufficient or there exists other reasons that make it difficult to identify this type of entities.

[4]http://www.medis.or.jp

6 Conclusions

This paper described our systems to extract personal and medical information from medical texts. We implemented a simple system based on structured perceptron as a first step toward more effective Japanese medical text processing systems, and extended it to systems based on another machine learning algorithm and on a morphological analyzer with a domain-specific dictionary. Moreover, we analyzed its performance and issues for achieving the goal. The result on the MedNLP dataset indicates that classification of medical entities into their modality classes, especially the suspicion class, is difficult. However, our analysis revealed that the terms and expressions in medical texts have useful patterns and characteristics that could be exploited for more accurate extraction.

Although it found that it was not very effective to use output of the morphological analyzer with domain-specific dictionary, we are aiming to use knowledge resources in more effective ways, e.g. incorporating dictionary features into classifiers. Additionally, we plan to explore more useful features such as suffix and prefix features for development of more advanced systems.

References

M. Collins. 2002. Discriminative training methods for hidden Markov models: Theory and experiments with perceptron algorithms. In *Proceedings of the 2002 conference on empirical methods in natural language processing (EMNLP)*, pages 1–8.

K. Crammer, O. Dekel, J. Keshet, S. Shalev-Shwartz, and Y. Singer. 2006. Online passive-aggressive algorithms. *The Journal of machine learning research (JMLR)*, 7:551–585.

J.R. Finkel, T. Grenager, and C. Manning. 2005. Incorporating non-local information into information extraction systems by gibbs sampling. In *Proceedings of the 43rd annual meeting on association for computational linguistics (ACL)*, pages 363–370.

S. Higashiyama, M. Blondel, K. Seki, and K. Uehara. (to appear). Named entity recognition exploiting category hierarchy using structured perceptron. *IPSJ Transactions on mathematical modeling and its applications.*

R. Johansson and P. Nugues. 2008. Dependency-based semantic role labeling of propbank. In *Proceedings of the 2008 conference on empirical methods in natural language processing (EMNLP)*, pages 69–78.

S Kaneko, N Fujita, Y Ugawa, T Kawamoto, H Takeuchi, M Takekoshi, and H Ohtake. 2003. Life science dictionary: a versatile electronic database of medical and biological terms". *Dictionaries and Language Learning: How can Dictionaries Help Human & Machine Learning?*, pages 434–439.

T Kudo, K Yamamoto, and Y Matsumoto. 2004. Applying conditional random fields to japanese morphological analysis. In *Proceedings of the 2004 conference on empirical methods in natural language processing (EMNLP)*, pages 230–237.

J. Lafferty, A. McCallum, and F. Pereira. 2001. Conditional random fields: probabilistic models for segmenting and labeling sequence data. In *Proceedings of the 18th international conference on machine learning (ICML)*, pages 282–289.

P. F Laquerre and C Malon. 2013. NECLA at the medical natural language processing pilot task (MedNLP). In *Proceedings of the 10th NTCIR conference*, pages 725–727.

A. McCallum and L. Wei. 2003. Early results for named entity recognition with conditional random fields, feature induction and web-enhanced lexicons. In *Proceedings of the 7th conference on natural language learning (CoNLL-2003)*, pages 188–191.

Y Miura, T Ohkuma, H Masuichi, E Yamada, E Aramaki, and K Ohe. 2013. UT-FX at NTCIR-10 MedNLP: incorporating medical knowledge to enhance medical information extraction. In *Proceedings of the 10th NTCIR conference*, pages 728–731.

M. Morita, Y. Kano, T. Ohkuma, M. Miyabe, and E. Aramaki. 2013. Overview of the NTCIR-10 MedNLP task. In *Proceedings of the 10th NTCIR conference*, pages 696–701.

B. Settles. 2004. Biomedical named entity recognition using conditional random fields and rich feature sets. In *Proceedings of the International Joint Workshop on Natural Language Processing in Biomedicine and its Applications (JNLPBA)*, pages 104–107.

F. Sha and F. Pereira. 2003. Shallow parsing with conditional random fields. In *Proceedings of the 2003 Conference of the North American Chapter of the Association for Computational Linguistics on Human Language (HLT-NAACL)*, pages 213–220.

Hyun-Je Song, Jeong-Woo Son, Tae-Gil Noh, Seong-Bae Park, and Sang-Jo Lee. 2012. A cost sensitive part-of-speech tagging: differentiating serious errors from minor errors. In *Proceedings of the 50th Annual Meeting of the Association for Computational Linguistics (ACL)*, pages 1025–1034.

Towards High-Reliability Speech Translation in the Medical Domain

Graham Neubig[1], **Sakriani Sakti**[1], **Tomoki Toda**[1], **Satoshi Nakamura**[1],
Yuji Matsumoto[1], **Ryosuke Isotani**[2], **Yukichi Ikeda**[2]

[1] Graduate School of Information Science, Nara Institute of Science and Technology
8916-5 Takayama-cho, Ikoma-shi, Nara, Japan
{neubig,ssakti,tomoki,s-nakamura,matsu}@is.naist.jp

[2] NEC Corporation
5-7-1 Shiba, Minato-ku, Tokyo, Japan
{r-isotani@bp,y-ikeda@df}.jp.nec.com

Abstract

In this paper, we describe the overall design for a speech translation system that aims to reduce the problems caused by language barriers in medical situations. As first steps to building a system according to this design, we describe a collection of a medical corpus, and some translation experiments performed on this corpus. As a result of the experiments, we find that the best of three modern translation systems is able to translate 33%-81% of the sentences in a way such that the main content is understandable.

1 Introduction

One of the most important elements to provision of high-quality medical service is communication between medical practitioners and patients. However, in situations where practitioners and patients do not share a common language, the language barrier prevents effective communication, making proper diagnosis and treatment much more difficult. Language barriers occur in medical situations with immigrants who may speak the language of their country of residence to some extent, but not enough to effectively communicate medical symptoms. There is also the case of medical tourism, where tourists may visit another country to receive high-quality or affordable medical treatment that is not available in their home country.

One potential method for overcoming the communication barrier in medical situations is through the use of automatic speech translation technology (Nakamura, 2009). Automatic translation of speech in medical situations can be expected to be challenging for a number of reasons. The first reason is that communication of incomplete or incorrect information could lead to a mistaken diagnosis with severe consequences, and thus extremely high levels of *accuracy* and *reliability* are required. The second reason is that conversation in the medical domain has its own unique vocabulary and expressions, and thus it is natural to assume that we must *adapt* the system appropriately to the medical domain.

There has been some previous work attempting to adapt communication technology to meet these two challenges. Eck et al. (2004) focus on adapting a translation system to medical vocabulary, although the focus on text translation of medical documents instead of speech translation for communication. Miyabe et al. (2007) propose a system for reliable multilingual communication, but rely on a graphical interface that is something like a powerful bilingual phrasebook adapted to communication at a hospital reception desk.

In this paper, we describe our vision for full speech translation in medical situations, and some first steps to achieve this vision. First, in Section 2 we describe our overall design for the speech translation system. This system includes the common components of automatic speech recognition (ASR), machine translation (MT), and text-to-speech (TTS), augmented to adapt each component to the task at hand. We also consider what is necessary to ensure the reliability of translation results, and consider the use of a system to allow the conversation to be forwarded to human medical interpreters when necessary.

In the first step towards achieving a translation system for the medical domain we have also collected a medical-domain corpus for Japanese-English and Japanese-Chinese translation, as described in Section 3. We share some insights gained in collecting this corpus, particularly comparing and contrasting text data from a medical domain bilingual phrasebook, and actual conversational data gathered during doctors' visits.

Based on this data, we then build several prototype translation systems for the four language pairs as described in Section 4. We perform au-

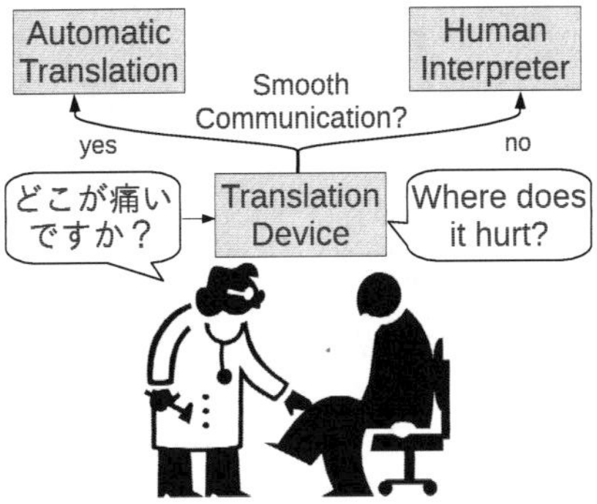

Figure 1: An overview of the use scenario for the medical translation system.

tomatic and manual evaluation of the results and evaluate how close we are to our goal of creating a system that can provide a first wave of assistance in medical situations. In particular, we find that over all four (relatively difficult) language pairs, we are on our way towards creating a practical medical machine translation system, with from 33%-81% of sentences over two tasks and four language pairs having all content understandable with some effort.

Finally, in Section 5 we conclude the paper with a discussion of future work.

2 Medical Translation System Design

We show the overall use scenario in Figure 1. In a typical doctor's visit, patient first enters the doctor's office, speaks with receptionists, and fills out forms. The patient will then enter the doctor's office and communicate with doctors/nurses. In order to introduce a speech translation system for use in this scenario, we will provide a device that will then translate between the language of the patient and that of the medical practitioners.

The device provides two possible methods of communication. The first is through the use of automatic speech translation technology, where the speaker's voice will be recorded, recognized, translated, and synthesized entirely automatically. In addition, as speech translation technology is still far from perfect, the device will also have the ability to connect to an actual human medical interpreter located in a call center. However, as the cost of hiring and maintaining medical interpreters is quite high, we would also like to reduce our reliance on human effort as much as possible. Thus, each device will use automatic translation by default, but also have functionality to connect to human interpreters, either based on a manual request of one of the users, or through automatic detection of when the dialogue is going poorly, such as the method described by Walker et al. (2000).

Even with this fall-back to human interpreters, it is still desirable that the automatic translation system is effective as possible. In order to ensure this, we must be certain that the ASR, MT, and TTS models are all tuned to work as well as possible in medical situations. Some potential problems that we have identified so far based on our analysis of data are as follows:

Specialized Vocabulary: Perhaps the most obvious problem is that the ASR, MT, and TTS systems must all be able to handle the specialized vocabulary and usage that occurs in the medical domain. For example, most we found that unadapted systems had trouble handling specialized terms such as "cardiogram," medicine names such as "Sudafed," and disease names such as "chicken pox." This will require the creation of domain specific corpora/dictionaries, and domain adaptation for each of the components (Leggetter and Woodland, 1995; Bellegarda, 2004).

Conversational Speech: The speech during doctor's visits will generally be somewhat informal and conversational when compared to that of speeches, news, or other more formal locations. As a result, we can expect ASR to be more difficult due to fillers, disfluencies and other factors (Goldwater et al., 2010).

Translation/Synthesis of Erroneous Input: As we can expect ASR not to be perfect, it will be necessary to be able to translate input that contains errors. This problem can potentially be ameliorated by passing multiple speech recognition hypotheses to translation (Ney, 1999), and jointly optimizing the parameters of ASR and MT (Zhang et al., 2004; Ohgushi et al., 2013). In addition, it will also be necessary to resolve difficulties in TTS due to grammatical errors, lack of punctuation, and unknown words (Parlikar et al., 2010).

While all of these problems need to be solved to provide high-reliability speech translation systems, in this paper as a first step we focus mainly on the MT system, and relegate the last problem of integration with ASR to future work.

3 Medical Translation Corpus Construction and Analysis

In this section, we describe our collection of a tri-lingual (Japanese, English, Chinese) corpus to serve as an initial testbed for our medical translation experiments, and an analysis of the corpus.

3.1 Corpus Construction

In general, when creating a corpus for training/testing a machine translation system, it is important to collect content that is as close as possible to that which we will encounter when the system is actually used. In our medical translation situation, this is true for both vocabulary (the corpus must cover special medical terms) and for speaking style (the corpus must have a similar style to that used by actual doctors and patients speaking through the system). There is also the practical concern that the cost of corpus collection is high, so we would like to perform collection in efficient a manner as possible.

Based on these principles, we designed and collected the following two corpora:

Medical Phrasebooks: The first corpus consists of sentences designed based on sentences from Japanese-English bilingual phrasebooks designed for interpreters focusing on the medical domains. Chinese translations were obtained by translating each phrase from Japanese to Chinese. This corpus has the advantages of relatively efficient construction, and good coverage of medical-domain terminology, but the conversations are not necessarily exactly representative of the conversations that actually occur at a doctor's office.

Medical Conversation: The second corpus we gathered consists of actual conversations between the patient and the receptionists or doctors recorded during a doctor's visit. The doctors and receptionists were all actual practicioners, but for privacy reasons the person acting as a patient was actually healthy, but given a scenario to act out. Conversations were recorded in Japanese and all participants were native Japanese speakers. The conversations were then segmented by utterance and translated into English and Chinese. This corpus has the advantage of being highly natural and covering medical domain terminology, but requires a large amount of time and effort for the creation of scenarios, gathering the participants, execution of the actual dialog, and transcription/translation of the results.

			Word		
		Sent.	ja	en	zh
Phrase	Train	3420	68k	43k	38k
	Dev	855	17k	12k	9.6k
	Test	855	17k	12k	9.6k
Conv.	Train	671	5.6k	4.7k	3.4k
	Dev	168	1.4k	1.3k	900
	Test	168	1.5k	1.2k	880

Table 1: Size in sentences and words of each language for each split for the phrasebook and conversation corpora.

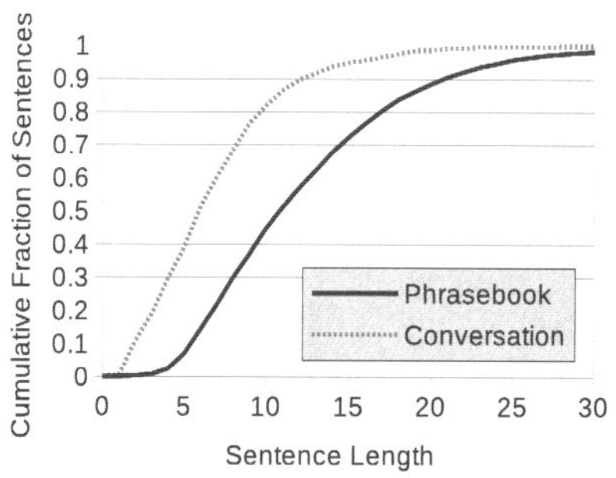

Figure 2: The cumulative length distribution of sentences in each corpus.

At the end of the collection, we had 5130 and 1007 sentences for the phrasebook and conversation corpus respectively. In addition, we create three splits of the corpus for use in the training, tuning, testing of our machine translation system with a ratio of 4:1:1. The final size of the data in all of these corpora is shown in Table 1.

3.2 Corpus Analysis

In this section, we describe some insights gained from the analysis of both corpora, with some examples illustrating the features of each corpus in Table 2.

One feature of the data with major implications is that there were large differences in speaking style between the phrasebook and conversation corpora. The data in the phrasebook corpus generally consisted of longer sentences, while the

Phrase	1)	I was sewing my jeans using a sewing machine and the needle broke and stabbed my left cheek.
	2)	I have been told that I have early indications of liver cirrhosis.
Conv.	1)	No more than two vials of blood. Possibly three, if for a blood sugar test.
	2)	Go straight, and on your left there is a green chair. / Here? Which way should I face?

Table 2: Examples of sentences (or several sentences separated by slashes) from the phrasebook and conversational corpora.

majority of the utterances in the conversation corpus contained short questions, requests, responses, and commands. This trend of longer sentences in the phrasebook corpus is shown clearly in Figure 2, which shows the cumulative length distribution of English sentences under a certain length in both corpora. Focusing on sentences under length 15, we can see that this covers a total of 95% of the conversation corpus, but only 72% of the phrasebook corpus.

In addition, the language in the conversation corpus is significantly less formal, particularly in Japanese where spoken language includes features such as dropped subjects or particles and abbreviations, which rarely occur in written language (Neubig et al., 2012). We hypothesize that in a cross-lingual medical conversation situation, the content of the utterances will fall somewhere between these two situations, as the content will be conversational, but the kind of natural and informal interaction seen two native speakers will be difficult to achieve through an automatic translation system.

The second enlightening feature of the two corpora that we noticed was that medical terminology was significantly less prevalent in the conversation corpus. This is also natural, as actual patients to a doctors office will likely be unfamiliar with difficult medical terms, and thus the doctors will tend to explain in language that is understandable for their audience. This observation will likely carry over to computer-mediated medical communication as well. As a result, it is likely that adapting to medical terminology of the domain is somewhat less important than adapting to the conversational speaking style of the speech.

4 Preliminary Evaluation of Medical Machine Translation

In this section we describe a preliminary evaluation of the effectiveness of automatic translation on the medical domain data described in the previous section. In particular, we focus on the MT component, leaving evaluation of ASR, TTS, and the system as a whole for future work.

4.1 Experimental Setup

For the tuning and test data for our translation system, we use the data described in the previous section. For training, 4,000 sentences is not enough to build an accurate MT system, so we add several additional corpora for each language pair. For Japanese-English parallel training data, we add the Eijiro dictionary[1] and its accompanying sample sentences, the BTEC corpus(Takezawa et al., 2002), and Wikipedia data from the Kyoto Free Translation Task (Neubig, 2011), for a total of 1.33M parallel sentences and 1.97M dictionary entries. For Japanese-Chinese parallel training data, we add a dictionary extracted from Wikipedia's language links[2], the BTEC corpus, and TED talks (Cettolo et al., 2012) for a total of 519k sentences and 184k dictionary entries. In addition, we add monolingual from English GigaWord with 22.5M sentences and Chinese Wikipedia with 841k sentences.

We compare three different statistical translation methodologies: phrase-based MT (PBMT, (Koehn et al., 2003)), hierarchical phrase-based MT (Hiero, (Chiang, 2007)), and forest-to-string MT (F2S, (Mi et al., 2008)). The reason why we test these three methodologies is because the former two methodologies do not rely on syntactic analysis, and thus may be more robust to conversational input that is ill-formed and/or informal. On the other hand, using syntactic information has been shown to improve translation, particularly between language pairs with different syntactic structures such as those we are handling in our experiments. Thus it will be interesting to see which methodology can produce better results, and also if any difference in the effectiveness of

[1] http://eijiro.jp

[2] http://wikipedia.org

the methodologies will be seen between the two corpora.

For training the translation models and decoding, we use the Moses toolkit (Koehn et al., 2007) for PBMT and Hiero, and the Travatar (Neubig, 2013) toolkit for F2S with the default settings. For training language models, we use SRILM (Stolcke, 2002), training Kneser-Ney smoothed 5-gram models for each individual language model training corpus, and linearly interpolating these models to maximize likelihood on the tuning corpus.

For tokenization we use the Stanford Tokenizer/Segmenter for English and Chinese (Tseng et al., 2005), and the KyTea segmenter for Japanese (Neubig et al., 2011). For syntactic parsing in English and Chinese we use a modified version of the Egret parser,[3] and for Japanese we use the Eda parser (Flannery et al., 2011) and the dependency-to-CFG conversion rules in the Travatar toolkit. Alignment is performed using the unsupervised aligner GIZA++ (Och and Ney, 2003) for Japanese-Chinese, and the supervised aligner Nile for Japanese-English (Riesa and Marcu, 2010), with the alignment models being trained on the alignments distributed with the Kyoto Free Translation Task.[4]

To measure translation accuracy, we use the automatic evaluation measures of BLEU (Papineni et al., 2002) and RIBES (Isozaki et al., 2010) measured over all sentences in the test corpus. We also perform a manual evaluation on 120 sentences from the phrasebook corpus and 80 sentences from the conversation corpus. These were randomly selected from all sentences of length 1-30, and graded using 1-5 adequacy (Goto et al., 2011) as our evaluation measure. We also report the percentage of sentences that received a rating of greater than or equal to 2, indicating that the main points of the sentence can be understood, possibly with some difficulty.

4.2 Experimental Results

The results of the experimental evaluation are shown in Figure 3. This graph shows many results, but we first focus on the furthestmost right graph, which shows the percentage of sentences understandable to some extent for each of the systems. From this graph, we can see that the scores range from 81% understandable sentences for Japanese-Chinese phrasebook sentences, to only 33% understandable sentences on Japanese-English phrasebook sentences. On the other hand, for conversational sentences, most language pairs hovered at around 55% understandable, with Japanese-English being significantly worse.

An in-depth analysis of the mistaken sentences identified several issues for improvement that were generally shared by all three systems.

Omitted pronouns: Japanese is a pro-drop language, which means that pronouns, usually the subject of the sentence can be omitted and inferred from the context. This phenomenon is particularly prevalent in the types of dialogue contained in the conversation corpus, with the majority of sentences having their subject omitted. Given that it is difficult for the translation systems used in the experiments to accurately reproduce these omitted subjects in a non-pro-drop target language such as English, it is likely that replacing these subjects in a preprocessing step would lead to gains in accuracy (Taira et al., 2012)

Dropped words: There were many cases where words central to the sentence were missing from the translation output by the system. This problem is rooted in a number of problems, such as words being mistakenly unaligned in the training data.

Word segmentation: Both Chinese and Japanese require the segmentation of raw text into words, but occasionally word segmentation errors occurred either due to conversational speech or specialized medical terms. Thus, using domain adaptation techniques (Neubig et al., 2011) to fix the word segmentations in the medical domain could potentially improve down-stream accuracy of translation as well.

Medical domain terms: As expected, there were a few medical domain terms not covered by corpora from more general domains, such as "Benadryl." However, the number was also relatively small, with only 5 untranslatable words occurring in a 200 sentence corpus.

Overall, an interesting shared point between the majority of members of the list is that they are not

[3] `http://github.com/neubig/egret`

[4] This preprocessing pipeline is available as part of the Travatar toolkit: `http://phontron.com/travatar/preprocessing.html`

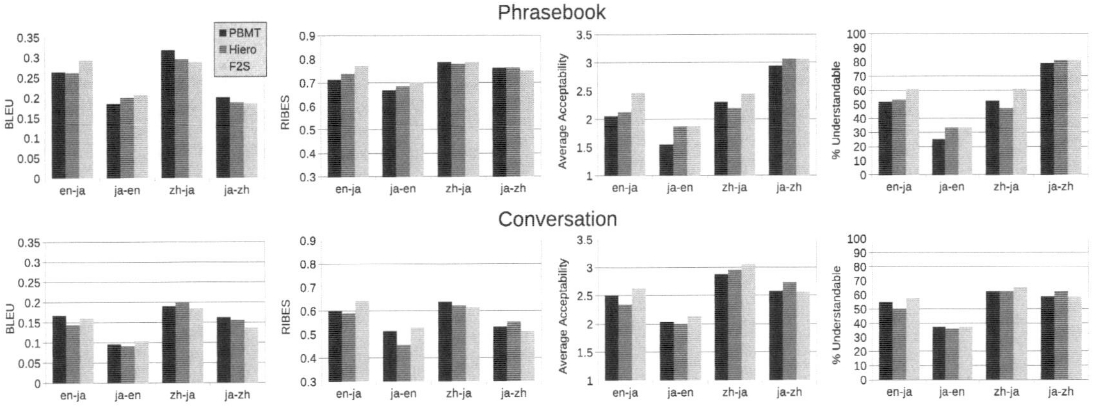

Figure 3: Results in BLEU, RIBES, Average Acceptability, and % Understandable for phrase-based (PBMT), hierarchical phrase based (Hiero), and forest-to-string (F2S) systems over translation of medical phrasebooks and conversations.

	Input	糖尿病でインスリンを使用しています。
	Ref	I use insulin because I have diabetes.
1	PBMT	I have diabetes using insulin.
	Hiero	Diabetes has been using insulin?
	F2S	I have been using insulin for diabetes.
	Input	反対の目をお願いします。
	Ref	Let's test the other eye.
2	PBMT	Other eye, please.
	Hiero	Please check your other eye.
	F2S	I'd like other eye.
	Input	もう一回眼をぱちぱちしてください。
	Ref	Once again, open and close your eyes.
3	PBMT	Their eyes again, please.
	Hiero	Their eyes again, please.
	F2S	Please ぱちぱち eye again.

Table 3: Examples of translations generated by each system for Japanese-English.

specific to medical translation, but more related to the style of the text. Thus while raising the level of medical MT will certainly involve covering medical terminology, it is also equally, if not more, important to overcome obstacles facing the more general speech translation task as well.

Finally, in Table 3, we show concrete examples for each of the three translation methods in Japanese-English translation. The first example is from the phrasebook data, uses some medical terms, and has a very typical syntactic structure for a written Japanese sentence. As a result F2S is able to translate almost perfectly, but PBMT and Hiero have reordering problems garbling the meaning of the sentence. The second example literally means "other eye, please," and PBMT is able to generate this very literal translation. Hiero, on the other hand, mistakenly makes the listener the subject of "check," and F2S mistakenly translates "please" as "I'd like," which doesn't make sense in this context. In the third example, all three systems have trouble translating the colloquial word for "blink one's eyes," with PBMT and Hiero dropping the word altogether, and F2S leaving it untranslated.

5 Conclusion and Future Work

In this paper, we described an overall design for a speech translation system that aims to reduce the problems caused by language barriers in medical situations. We describe a collection of a medical corpus, and some translation experiments performed on this corpus. As a result of the experiments, we find that the best of three modern translation systems is able to translate 33%-81% of the sentences in a way such that the main content is understandable.

While these preliminary results are encouraging, this is just the first step towards a full medical speech translation system. As described, there are still a number of challenges related to the MT module itself, including the handling of informal speech. These will further be compounded when combined with the need for robust ASR and TTS. However, given the potential for speech translation technology to be useful in medical situations, we believe that meeting these research challenges is a worthy target for research and development in the near future.

Acknowledgments

Part of this research was executed under the Commissioned Research for "Research and Development on Medical Communication Support System for Asian Languages based on Knowledge and Language Grid" of National Institute of Information and Communications Technology (NICT), Japan.

References

Jerome R Bellegarda. 2004. Statistical language model adaptation: review and perspectives. *Speech communication*, 42(1):93–108.

Mauro Cettolo, Christian Girardi, and Marcello Federico. 2012. WIT3: web inventory of transcribed and translated talks. pages 261–268.

David Chiang. 2007. Hierarchical phrase-based translation. *Computational Linguistics*, 33(2).

Matthias Eck, Stephan Vogel, and Alex Waibel. 2004. Improving statistical machine translation in the medical domain using the unified medical language system. In *Proc. COLING*, pages 792–798.

Daniel Flannery, Yusuke Miyao, Graham Neubig, and Shinsuke Mori. 2011. Training dependency parsers from partially annotated corpora. In *Proc. IJCNLP*, pages 776–784, Chiang Mai, Thailand, November.

Sharon Goldwater, Daniel Jurafsky, and Christopher D. Manning. 2010. Which words are hard to recognize? prosodic, lexical, and disfluency factors that increase speech recognition error rates. *Speech Communication*, 52(3):181–200.

Isao Goto, Bin Lu, Ka Po Chow, Eiichiro Sumita, and Benjamin K. Tsou. 2011. Overview of the patent machine translation task at the NTCIR-9 workshop. In *Proceedings of NTCIR*, volume 9, pages 559–578.

Hideki Isozaki, Tsutomu Hirao, Kevin Duh, Katsuhito Sudoh, and Hajime Tsukada. 2010. Automatic evaluation of translation quality for distant language pairs. In *Proc. EMNLP*, pages 944–952.

Phillip Koehn, Franz Josef Och, and Daniel Marcu. 2003. Statistical phrase-based translation. In *Proc. HLT*, pages 48–54, Edmonton, Canada.

Philipp Koehn, Hieu Hoang, Alexandra Birch, Chris Callison-Burch, Marcello Federico, Nicola Bertoldi, Brooke Cowan, Wade Shen, Christine Moran, Richard Zens, Chris Dyer, Ondrej Bojar, Alexandra Constantin, and Evan Herbst. 2007. Moses: Open source toolkit for statistical machine translation. In *Proc. ACL*, pages 177–180, Prague, Czech Republic.

Christopher J Leggetter and Philip C Woodland. 1995. Maximum likelihood linear regression for speaker adaptation of continuous density hidden markov models. *Computer Speech & Language*, 9(2):171–185.

Haitao Mi, Liang Huang, and Qun Liu. 2008. Forest-based translation. In *Proc. ACL*, pages 192–199.

Mai Miyabe, Kunikazu Fujii, Tomohiro Shigenobu, and Takashi Yoshino. 2007. Parallel-text based support system for intercultural communication at medical receptions. In *Intercultural Collaboration*, pages 182–192. Springer.

Satoshi Nakamura. 2009. Overcoming the language barrier with speech translation technology. *NISTEP Quarterly Review*, (31).

Graham Neubig, Yosuke Nakata, and Shinsuke Mori. 2011. Pointwise prediction for robust, adaptable Japanese morphological analysis. In *Proc. ACL*, pages 529–533, Portland, USA, June.

Graham Neubig, Yuya Akita, Shinsuke Mori, and Tatsuya Kawahara. 2012. A monotonic statistical machine translation approach to speaking style transformation. *Computer Speech and Language*, 26(5):349–370, October.

Graham Neubig. 2011. The Kyoto free translation task. `http://www.phontron.com/kftt`.

Graham Neubig. 2013. Travatar: A forest-to-string machine translation engine based on tree transducers. In *Proc. ACL Demo Track*, Sofia, Bulgaria, August.

Hermann Ney. 1999. Speech translation: Coupling of recognition and translation. In *Proc. ICASSP*, pages 517–520.

Franz Josef Och and Hermann Ney. 2003. A systematic comparison of various statistical alignment models. *Computational Linguistics*, 29(1):19–51.

Masaya Ohgushi, Graham Neubig, Sakriani Sakti, Tomoki Toda, and Satoshi Nakamura. 2013. An empirical comparison of joint optimization techniques for speech translation. In *Proc. 14th InterSpeech*.

Kishore Papineni, Salim Roukos, Todd Ward, and Wei-Jing Zhu. 2002. BLEU: a method for automatic evaluation of machine translation. In *Proc. ACL*, pages 311–318, Philadelphia, USA.

Alok Parlikar, Alan W Black, and Stephan Vogel. 2010. Improving speech synthesis of machine translation output. In *Proc. 11th InterSpeech*, pages 194–197.

Jason Riesa and Daniel Marcu. 2010. Hierarchical search for word alignment. In *Proc. ACL*, pages 157–166.

Andreas Stolcke. 2002. SRILM - an extensible language modeling toolkit. In *Proc. 7th International Conference on Speech and Language Processing (ICSLP)*.

Hirotoshi Taira, Katsuhito Sudoh, and Masaaki Nagata. 2012. Zero pronoun resolution can improve the quality of J-E translation. In *Proceedings of the Sixth Workshop on Syntax, Semantics and Structure in Statistical Translation*, pages 111–118, Jeju, Republic of Korea, July.

Toshiyuki Takezawa, Eiichiro Sumita, Fumiaki Sugaya, Hirofumi Yamamoto, and Seiichi Yamamoto. 2002. Toward a broad-coverage bilingual corpus for speech translation of travel conversations in the real world. In *Proc. LREC*, pages 147–152.

Huihsin Tseng, Pichuan Chang, Galen Andrew, Daniel Jurafsky, and Christopher Manning. 2005. A conditional random field word segmenter for sighan bakeoff 2005. In *Proceedings of the Fourth SIGHAN Workshop on Chinese Language Processing*. Jeju Island, Korea.

Marilyn Walker, Irene Langkilde, Jerry Wright, Allen Gorin, and Diane Litman. 2000. Learning to predict problematic situations in a spoken dialogue system: experiments with how may I help you? In *Proc. 6th Conference on Applied Natural Language Processing*, pages 210–217.

Ruiqiang Zhang, Genichiro Kikui, Hirofumi Yamamoto, Taro Watanabe, Frank Soong, and Wai Kit Lo. 2004. A unified approach in speech-to-speech translation: integrating features of speech recognition and machine translation. In *Proc. COLING*, pages 1168–1174.

Finding Every Medical Terms by Life Science Dictionary for MedNLP

Shuji Kaneko
Graduate School of Pharmaceutical Sciences, Kyoto University
Kyoto, Japan
skaneko@pharm.kyoto-u.ac.jp

Nobuyuki Fujita
National Institute of Technology and Evaluation
Tokyo, Japan
fujitan@nifty.com

Hiroshi Ohtake
Center for Arts and Sciences
Fukui Prefectural University
Fukui, Japan
ohtake@fpu.ac.jp

ABSTRACT

We have been developing an English-Japanese thesaurus of medical terms for 20 years. The thesaurus is compatible with MeSH (Medical Subject Headings developed by National Library of Medicine, USA) and contains approximately 30 thousand headings with 200 thousand synonyms (consisting of the names of anatomical concepts, biological organisms, chemical compounds, methods, disease and symptoms). In this study, we aimed to extract medical terms as many as possible from the test data by a simple longest-matching Perl script. After changing the given UTF-8 text to EUC format, the matching process required only 2 minutes including loading of a 10 MB dictionary into memory space with a desktop computer (Apple Mac Pro). From the 0.1 MB test document, 2,569 terms (including English spellings) were tagged and visualized in a color HTML format. Particularly focusing on the names of disease and symptoms, 893 terms were found with several mistakes and missings. However, this process has a limitation in assigning ambiguous abbreviations and misspelled words. The simple longest-matching strategy may be useful as a preprocessing of medical reports.

Keywords
Life Science Dictionary, Medical Thesaurus, MeSH

Team Name
LSDP (standing for Life Science Dictionary Project)

Subtask
Free Task (finding every medical terms)

1. INTRODUCTION
The Life Science Dictionary (LSD) project, founded in 1993, is a research project by us to develop a systematic database for life science (of course, including medical) terms and tools for the convenience of life scientists [1]. Our services are designed to provide and encourage access within the scientific community to the most up-to-date and comprehensive information on English-Japanese translation dictionary of life science terms. In keeping with the users' expectations, we have been enriching and refining the database records to a medical thesaurus compatible with MeSH (Medical Subject Headings developed by National Library of Medicine, USA) thesaurus. Recent version of LSD contains approximately 30 thousand headings with 200 thousand English and Japanese synonyms, consisting of the names of anatomical concepts, biological organisms, chemical compounds, methods, disease and symptoms.

One of the practical applications of thesaurus is text mining. For example, adverse drug events can be rapidly extracted by finding the causal relationship of drug treatment and related symptoms recorded in medical records. Favorably, our thesaurus contains a wide range of medical concepts as mentioned. In addition, we have previously developed a series of gloss-embedding Perl scripts for medical English texts [2]. In this study, therefore we aimed to tag every medical term (Japanese and English) as many as possible to evaluate the robustness of thesaurus and tagging program.

2. METHODS

2.1 Dictionary
A tagger dictionary was made from LSD database as an EUC text file, which contains approximately 200,000 rows and 4 columns: (1) synonym strings, (2) subject heading strings, (3) category of term, (4) subject heading ID (from MeSH). For the category of terms, all terms were classified and marked by one of the following categories according to the MeSH tree: anatomy, biological, disease, molecule, method, and knowledge (Fig. 1).

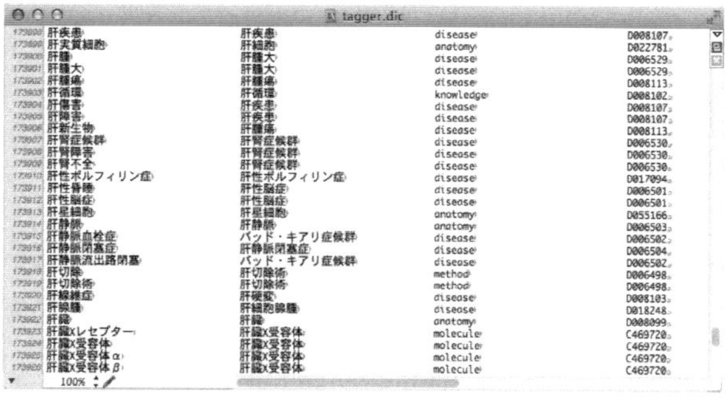

Fig. 1. Contents of tagger dictionary

2.2 Perl scripts
To take full advantage of the LSD in which many phrases have been registered, "the longest matches first" principle was adopted in the matching process. For this purpose, the tagger dictionary was sorted in the descending order of byte lengths, and text matching was performed for each of the dictionary entries in this order.

For the sake of the speed of text matching in Perl language, both the text and the dictionary were first converted to EUC encoding, and they were treated as byte strings in the matching process. Also, all two-byte roman characters were converted to corresponding ASCII characters, and multi-byte characters unique in Unicode were converted to appropriate ASCII character(s) as far as possible.

For better readability of the resulting data as well as for the ease of any secondary use, a standard HTML format was used as the output in which unique "class" attribute was assigned to each of the category (Fig. 2A). This allows the users to customize text coloring even after the output of the data. We also added a 'mouse-over heading' feature, in which the embedded subject heading of the term will be displayed when the cursor was placed over the tagged term (Fig. 2B). In addition, by clicking the tagged part, the user can confirm the thesaurus entry in our WebLSD online dictionary system.

A

B

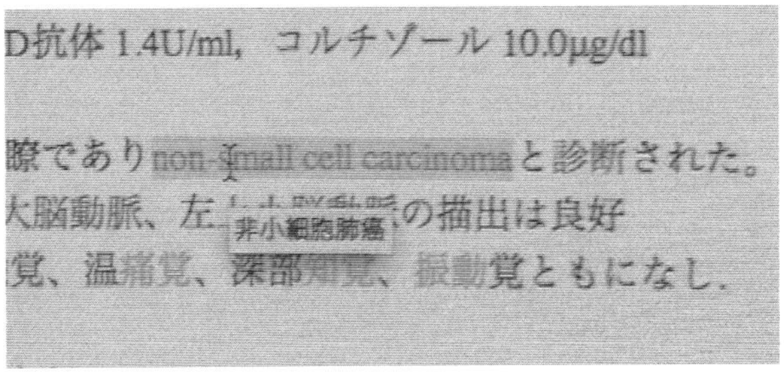

Fig. 2 HTML output (A) and mouse-over heading function (B)

3. RESULTS

3.1 Speed

For the test set containing 1,121 sentences, tagging process including UTF8-to-EUC conversion, 120 seconds were required with our Perl script by an Apple Mac Pro machine (3.2GHz Quad-Core Intel Xeon, 16GB memory). The speed of tagging seemed to be simply proportional to the length of the source text.

3.2 Overall result

From the 0.1 MB test document, 2,569 terms (including English spellings) were tagged and isolated. The most abundant category was the names of disease and symptoms, and 893 terms were found (Table 1).

Table 1. Number of tagged terms

Category	Tagged
Anatomy	439
Biological	35
Disease (or Symptom)	893
Molecule (or Drug)	395
Method (or Index)	622
Other knowledge	185
Total	2,569

3.3 Missed or incorrect tags

In addition to many correctly-tagged terms, several patterns of missed or incorrect tags were found.

The mostly missed terms were English abbreviations (Table 2). Especially, in the description of clinical test data, a variety of abbreviations were used, which cannot be marked. Since the meanings of 2- or 3-word abbreviations are ambiguous, we had omitted most of the abbreviations from tagger dictionary. However, if we know the part of document is apparently indicating clinical data, we can make a specific tagger dictionary for clinical tests. Similarly, some of the drug names were written in acronyms or non-universal abbreviations.

Table 2. List of missed abbreviations

Subcategory	Examples
Clinical test	T-Chol, Hb, Plt, eosino, BP, MPO, PaCO2, ALT, Cre, T-Bil, ZTT, APTT, etc.
Drug name	DIC（ダカルバジン）
	CLDM（クリンダマイシン）
	PIPC（ピペラシリン）
	PAPM/BP（パニペネム・ベタミプロン合剤）

The most typical pattern of incorrect tag was 'partly-tagged' term (Table 3). In these cases, part of unit concepts were registered in the dictionary, however, the combination of two or more concepts is common particularly in the names of disease and symptom, which were not completely covered in our thesaurus.

Table 3. Examples of partly-tagged words

Partial	Compounded	More complex case
温痛覚	Murphy 徴候	眼球の黄染
顔面紅斑	心音不整	前頚部の腫脹
日光過敏	眼球結膜黄染	胆嚢軽度腫大
剥離爪	肺 MAC 症	下肺には honey comb

3.4 Misspelling and typographical issue

To our surprise, there were many misspellings and typographical errors, even in Japanese terms, in the test document. Precise text matching did not tag incorrect spellings that medical doctor can recognize their meanings.

Table 4. List of misspellings

In the text	Correct
predonisolone	prednisolone
theophyline	theophylline
mycobacterium abcessus	Mycobacterium abscessus
Enterococcus fecalis	Enterococcus faecalis
Klebsiella pneumonoae	Klebsiella pneumoniae
コルトコフ音	コロトコフ音
グルドパ	グルトパ（Grtpa）
クオンテエンフエロン	クオンテイフエロン

4. DISCUSSION

With our tagging dictionary and scripts, most of medical terms were easily marked and visualized as a HTML document. From the 0.1 MB test document, 2,569 terms (including English spellings) were tagged and visualized in a color HTML format. Particularly focusing on the names of disease and symptoms, as much as 893 terms were found. Additional 'mouse-over heading' and web reference enables easy reviewing of the tagged terms.

Through this task, we have learnt the potential of our thesaurus and scripts in finding medical terms from given Japanese texts. However, this process has a limitation in assigning ambiguous abbreviations and misspelled words. Moreover, there is an insurmountable difficulty to accomplish a 'perfect matching' with a fixed text dictionary, since improvement of thesaurus is a laborious work. The simple tagging strategy may be useful as a preprocessing of medical reports. Combination of natural text processing with this tool will be convenient for the practical use.

5. REFERENCES

[1] Kaneko S, Fujita N, Ugawa Y, Kawamoto T, Takeuchi H, Takekoshi M, Ohtake H. 2003. Life Science Dictionary: a versatile electronic database of medical and biological terms. "Dictionaries and Language Learning: How can Dictionaries Help Human & Machine Learning", Asialex, pp.434-439.

[2] Ohtake H, Kawamoto T, Takekoshi M, Kunimura M, Morren B, Takeuchi H, Ugawa Y, Fujita N, Kaneko S. 2003. Development of a genre-specific electronic dictionary and automatic gloss-embedding system. "Dictionaries and Language Learning: How can Dictionaries Help Human & Machine Learning", Asialex, pp.445-449.

Proper and Efficient Treatment of Anaphora and Long-Distance Dependency: an Experiement with Medical Text

Wai Lok TAM[1], Yusuke MATSUBARA[1], Koiti HASIDA[1], Motoyuki TAKAAI[2], Eiji ARAMAKI[3], Mai MIYABE[3], and Hiroshi UOZAKI[4]

[1]Socia ICT Center, Graduate School of Information Science and Technology, The University of Tokyo
[2]Communication Technology Laboratory, Research and Technology Group, Fuji Xerox Co., Ltd.
[3]Design School, University of Kyoto
[4]Department of Pathology, School of Medicine, Teikyo University

1 Introduction

This paper is a follow-up to (Hasida et al.2012)'s work on pathological reports, a kind of medical text. Such reports have the following characteristics:

1. The composition of such reports has to follow a published set of strict guidelines (In our case, these guidelines are given in (JGCA2010). English version of these guidelines can be found in (JGCA2011).)
2. The subject matters of such reports are strictly limited to specimens submitted for pathological analysis.

These characteristics put the text in pathological reports under the category of controlled natural language, making it a better object text for semantic analysis and knowledge representation. Readers unfamiliar with controlled natural language are recommended to check the survey by (Schwitter2010).

The purpose of this paper is to present how to combine a CFG (Context-Free Grammar) with an ontology to account for both syntactic structures and semantic structures of sentences (and discourses) containing long-distance dependencies and anaphora found in pathological reports. The syntactic and semantic framework outlined in this paper are developed on the foundation of the Global Document Annotation (GDA) guidelines proposed by (Hasida2010).

When constructing our grammar, we have an application in mind. This application is auto-completion and hence speed matters. We want to do a bit more than bigrams can achieve with auto-completion such that the effect of an antecedent or a relative clause on user input can be captured. It is true that an elaborated feature structure-based grammars with hundreds of features would have little problem with anaphora and long distance dependencies. But speed is a problem for such a grammar. This leaves us with CFGs but typical CFGs can handle neither of the phenomena we are interested in. So we make CFGs do the job.

2 Components of Our Grammar

Essentially, our grammar works like a simple unification-based grammar. For illustrative purpose, let us first explain how it works as if it is a unification based grammar represented by directed acyclic graphs (DAGs). The nodes in one such graph are either concepts taken from an ontology or relations between them.

A baby version of our ontology trimmed down to a hierarchy of concepts relevant to examples sentences cited in this paper is given in figure 1.

(1) [$_{\text{N}}$ 小弯長] [$_{\text{NOADJ-GA}}$ 12cm] [, ,]
syouwantyou zyuni_senti COMMA
[$_{\text{N}}$ 大弯長] [$_{\text{NOADJ-GA}}$ 19.5cm] [。。]
daiwantyou zyukyutengo_senti PERIOD

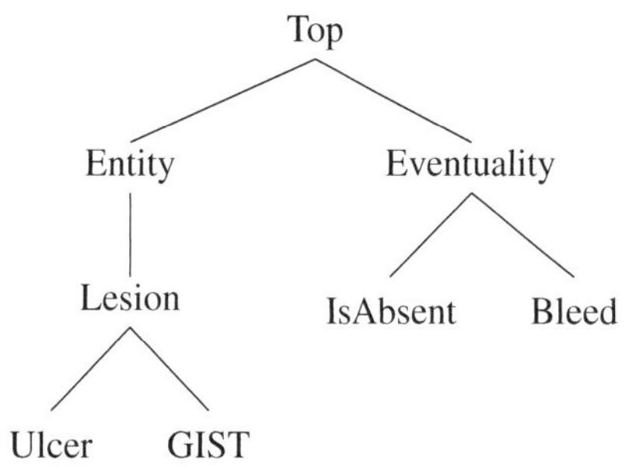

Figure 1: Hierarchy of Concepts Used in Example Sentences

Next comes the links between concepts in our ontology and words in our example sentences. These links are given in the lexicon illustrated by figure 2.

The links between concepts in our ontology and the meaning of a sentence are computed by a handful of semantic composition rules, the most fundamental of which is the rule for headed structure given in figure 3.

In figure 3, the mother(M), the head daughter (HD) and the nonhead daughter (ND) are determined by the combination of POS labels in the syntactic rule given below:

$$\text{S} \rightarrow \text{N}\phi \ \underline{\text{S-GA}}$$

The underlined daughter is the head daughter of the mother on the left hand side.

3 How the Components Work Together to Parse a Simple Sentence

Now let us parse an example sentence (1) with the syntactic rules. The parse tree is given in figure 4.

To make sense of the semantic composition going on here, some explanation for the path labels and the node

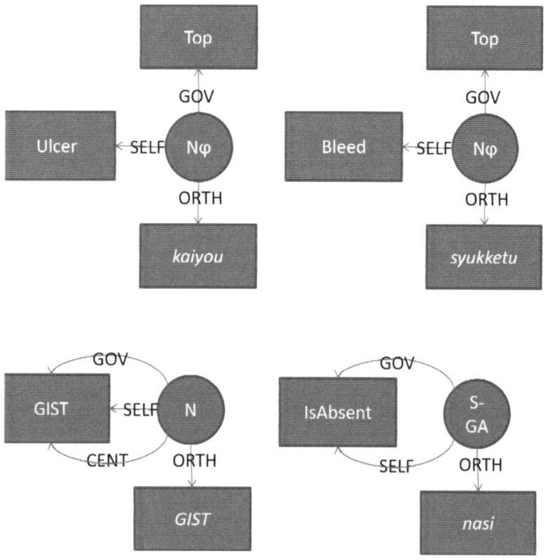

Figure 2: Lexicon

labels is probably needed. The path labels in upper cases SELF and GOV are fundamental to all constituents. The two ends of a SELF path are the P(art) O(f) S(peech) of a constituent and the meaning of the constituent. The two ends of a GOV path are the POS of a constituent and the meaning of the head on which the constituent depends. If a GOV path connects the same nodes as a SELF path, this means the constituent in question is not a dependent of any other constituent. The path label "theme" is a relation from the Top concept to the Abnormality concept defined in our ontology.

When the node labelled "*Top*" and connected by the GOV path to the Nϕ "*kaiyou*" unifies with the node labelled "*IsAbsent*" and connected to SELF path to the S-GA "*nasi*" as a result of the rule illustrated in figure 3, the "*IsAbsent*" concept is also specified as the domain of

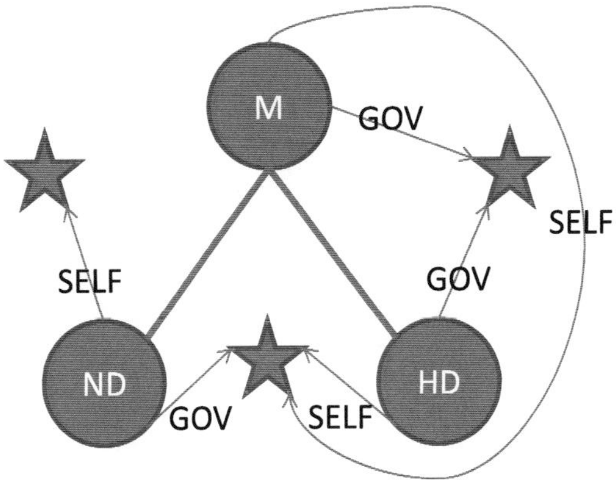

Figure 3: Semantic Rule for Headed Structures

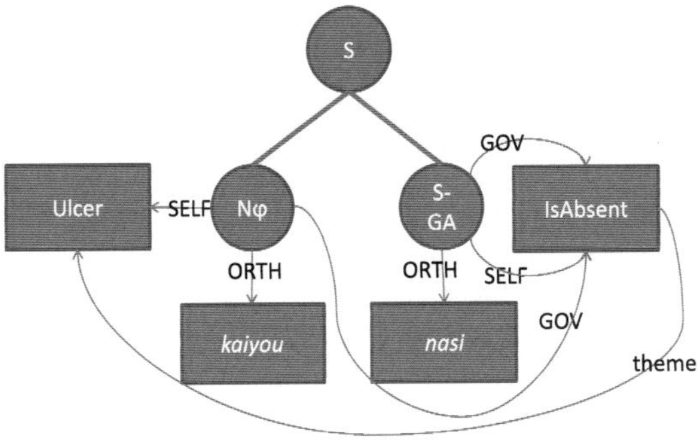

Figure 4: Parse Tree of an Example of Headed Structure

the "*theme*" relation. This way we connect the "*IsAbsent*" concept to the "*Ulcer*" concept by the "*theme*" relation, yielding the semantic representation of the example sentence "*Kaiyou Nasi*", which means "No ulcer is found".

4 Dealing with Anaphora

Now let us substitute the Nϕ in our example sentence with the verbal noun "*syukketu*", meaning "bleed". This introduces a gap in the sentence and the gap refers to the subject of "*syukketu*", which is nowhere to be found in the sentence. To resolve the zero anaphora, we need to store this gap somewhere. Meeting this need is one of the purposes of the CENT path we would like to introduce here. The CENT path also serves the need to pass up the value of the node it connects the POS node to such that both the antecedent and the anaphora can see each other. The percolation is handled by the rule illustrated in figure 5. Applying this rule to our example sentence yields the parse tree given in figure 6.

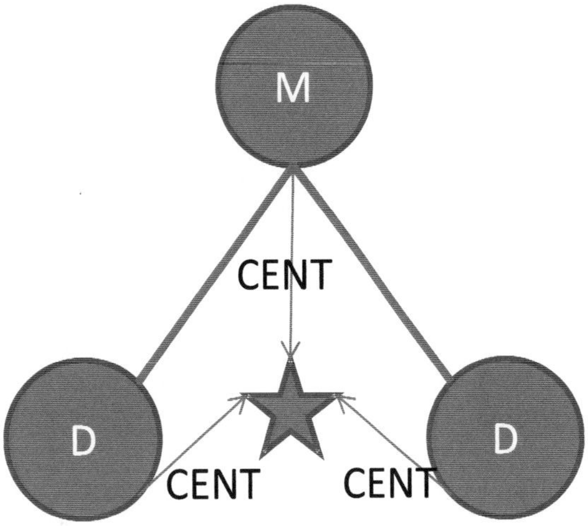

Figure 5: Percolation of Zero Anaphora

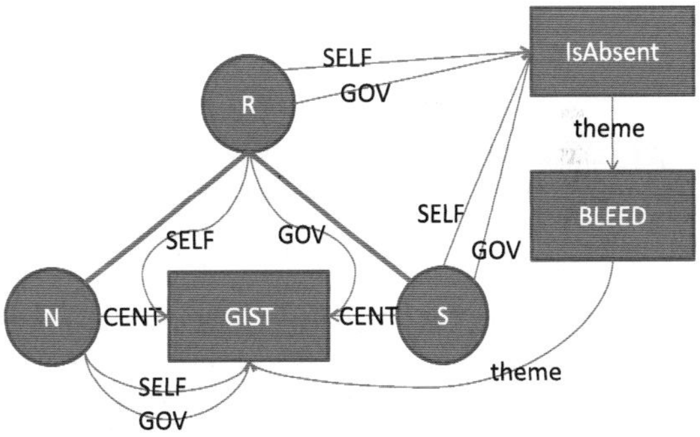

Figure 6: Parse Tree of an Example Containing an Anaphoric Expression

After passing the anaphoric gap to the root, we now come to the point to resolve the anaphora. In order to do this, we need an anaphora resolution rule, which is illustrated in figure 7 and another sentence containing the antecedent. Let us keep this sentence simple and make it constitute of a N(oun) "GIST", meaning "Gastrointestinal Stromal Tumor". When acting as an antecedent, the node connected by the CENT path to the POS node of it shares the same value with the node connected by the SELF path to the POS node, as illustrated in figure 2. We also need to add a syntactic rule for combining a N with a S to form a R(eport).

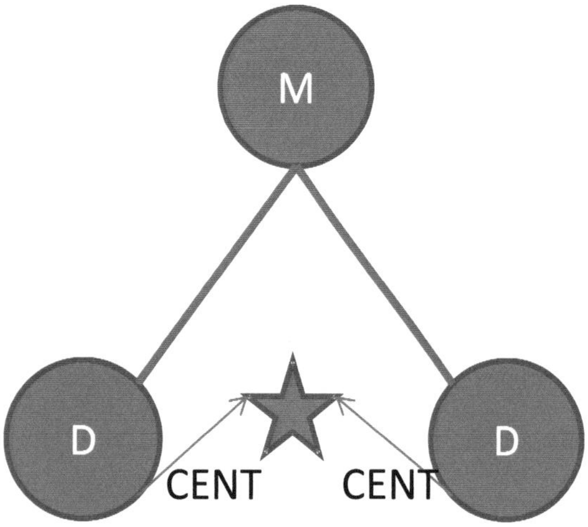

Figure 7: Resolving an Anaphora

$$\mathrm{R} \rightarrow \underline{\mathrm{N}}\ \underline{\mathrm{S}}$$

The semantic composition that goes hands in hand with this syntactic rule is illustrated in figure 8. Some parts of the semantic composition have nothing to do with anaphora resolution. They are there to make sure that the meaning of the mother R is the sum of the meaning of its parts. This is done by introducing two nodes connected by the SELF path to the R node and two nodes connected by the GOV node to the R node. So we have a pair of nodes for each daughter to pass up the values assigned to the pairs of nodes connected to it by its SELF path and GOV path. Putting everything together, we get the parse tree illustrated in figure 9.

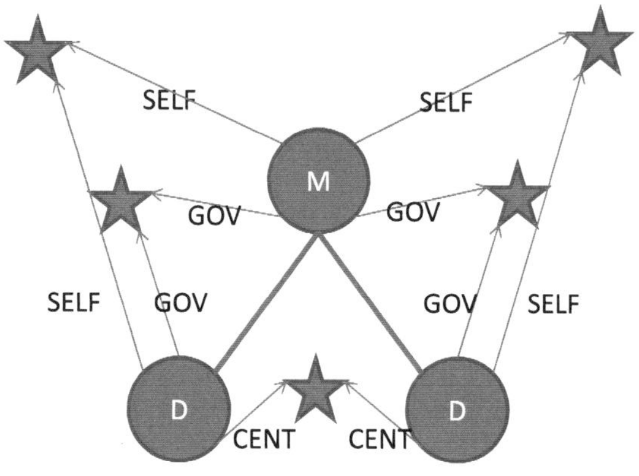

Figure 8: Conjoining Two Sentences and Resolving an Anaphora

5 Compiling DAGs into CFG rules

Now let us turn what we present so far into a CFG. The first step is to merge all nodes connected to the POS node by various paths into a single label in a predefined order such that the order can tell which value corresponds to which node. Assuming that the values are ordered: POS|CENT|SELF|GOV, the lexical entries illustrated in figure 2 are rewritten as:

$$\begin{array}{rcl} \mathrm{N}\phi|\_|Ulcer|Top & \rightarrow & "kaiyou" \\ \mathrm{N}\phi|\_|Bleed|Top & \rightarrow & "syukketu" \\ \mathrm{N}|GIST|GIST|GIST & \rightarrow & "GIST" \\ \text{S-GA}|\_|IsAbsent|IsAbsent & \rightarrow & "nasi" \end{array}$$

The two syntactic rules, which are typical CFG rules, would have to be rewritten as follows such that CFG rules can do the magic of semantic composition:

$$\begin{array}{rcl} \mathrm{S}|\_|IA|IA & \rightarrow & \mathrm{N}\phi|\_|U|T\ \ \underline{\text{S-GA}}|\_|IA|IA \\ \mathrm{S}|T|IA|IA & \rightarrow & \mathrm{N}\phi|T|B|T\ \ \underline{\text{S-GA}}|\_|IA|IA \\ \mathrm{R}|\_|G,IA|G,IA & \rightarrow & \mathrm{N}|G|G|G\ \ \mathrm{S}|T|IA|IA \end{array}$$

where "IA" stands for "*IsAbsent*", "U" stands for "*Ulcer*", "B" stands for "*Bleed*", "T" stands for "*Top*" and "G" stands for "*GIST*".

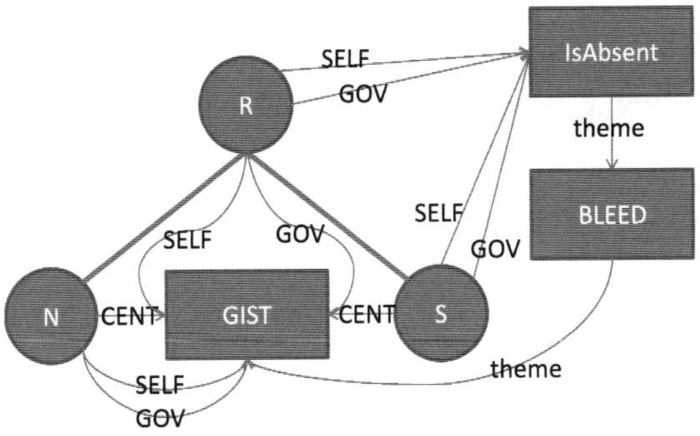

Figure 9: Parse Tree of a Report Containing an Anaphora and the Antecedent it refers to

When the lexicon contains only words referring to leaf concepts, this set generates the same set of sentences as the unification based grammar presented earlier on. If we add the N *"byouhen"*, which is assigned the superclass of GIST and Ulcer, Lesion as the meaning of it to the lexicon, the CFG given here becomes no longer equivalent to the unification based grammar presented in the beginning of this section. So we need to take the step of expanding the rules to cover words denoting ancestors or descendants of the CENT values, the SELF values and GOV values that make up parts of symbols in a CFG rule. This step adds the following rules to our CFG:

$$\begin{array}{rcl} \text{S}|\_|IA|IA & \rightarrow & \text{N}\phi|\_|L|T \;\; \underline{\text{S-GA}}|\_|IA|IA \\ \text{R}|\_|L,IA|L,IA & \rightarrow & \text{N}|L|L|L \;\; \text{S}|T|IA|IA \end{array}$$

where "L" stands for *"Lesion"*.

6 Conclusion and Future Work

The point of giving the details of compiling a unification based grammar into a CFG is to make the beauty of the small number of features to stand out. Assuming a rule with m features, each having n possible values, adding k values to an existing feature would increase the number of rules by $k \cdot (m-1) \cdot n$. Assigning k values to a new feature, would increase the number of rules by $k \cdot m \cdot n$. So a strictly limited number of features speed up things. This adds to the speed resulting from the lower complexity value $O(n^3)$ of a CFG when compared to the exponential complexity of a feature structure based grammar. The combined power of our grammar design and compilation into CFG make it possible for us to answer the needs of a real time task like auto-completion. Without any optimization, we achieve $500ms$ per sentence when running our grammar on a parser written in Python, an interpreted language. This is pretty much the best a deep parser can achieve as reported by (Matsuzaki et al.2007) on an older but likely to be faster machine than ours, a notebook computer with a 2.40Ghz Core 2 Duo processor. We are hopeful that we can further improve our speed by simply implementing our parser in a compiled language. When it is done, our grammar is expected to support auto-completion of less controlled natural languages such as the language used in nursing reports.

References

Koiti Hasida, Wailok Tam, Taiichi Hashimoto, Motoyuki Takaai, and Eiji Aramaki. 2012. Ontoroji taiou bunpou riron to sono kousokushori no tame no conpaireson. In *Proceedings of Gengo Shori Gakkai*, Hiroshima, Japan.

Koiti Hasida. 2010. Global document annotation. `http://i-content.org/GDA`.

Japanese Gastric Cancer Association JGCA, editor. 2010. *Japanese Classification of Gastric Carcinoma*. Kanehara.

Japanese Gastric Cancer Association JGCA. 2011. Japanese classification of gastric carcinoma: 3rd english edition. In *Gastric Cancer*, pages 101–112. Springer.

Takuya Matsuzaki, Yusuke Miyao, and Junichi Tsujii. 2007. Efficient hpsg parsing with supertagging and cfg-filtering. In *Proceedings of IJCAI2007*.

Rolf Schwitter. 2010. Controlled natural languages for knowledge representation. In *Proceedings of the 23rd International Conference on Computational Linguistics*, pages 1113–1121.

A Comparison of Rule-Based and Machine Learning Methods for Medical Information Extraction

Osamu Imaichi, Toshihiko Yanase, and Yoshiki Niwa
Hitachi, Ltd., Central Research Laboratory
1-280, Higashi-Koigakubo
Kokubunji-shi, Tokyo 185-8601
{osamu.imaichi.xc,toshihiko.yanase.gm,yoshiki.niwa.tx}@hitachi.com

Abstract

This year's MedNLP (Morita and Kano, et al., 2013) has two tasks: de-identification and complaint and diagnosis. We tested both machine learning based methods and an ad-hoc rule-based method for the two tasks. For the de-identification task, the rule-based method achieved slightly higher results, while for the complaint and diagnosis task, the machine learning based method had much higher recalls and overall scores. These results suggest that these methods should be applied selectively depending on the nature of the information to be extracted, that is to say, whether it can be easily patternized or not.

1 Introduction

Machine learning based and rule-based methods are the two major approaches for extracting useful information from natural language texts. To clarify the pros and cons of these two approaches, we applied both approaches to this year's MedNLP tasks: de-identification and complaint and diagnosis.

For the de-identification task, ages and times, for example, are seemingly a type of information that can be patternized quite easily. In such cases, an ad-hoc rule-based method is expected to perform relatively well. In contrast, the complaint and diagnosis task would seem to have much more difficulty patternizing information, so a machine learning approach is expected to provide an effective methodology for tackling these problems.

2 Machine Learning Approach

In this section, we explain how the machine learning based approach works.

2.1 Sequential Labeling by using CRF

We formalized the information extraction task as a sequential labeling problem. A conditional random field (CRF) (Lafferty and McCallum, et al., 2001) was used as the learning algorithm. We used CRFsuite[1], which is an implementation of first-order linear chain CRF.

The CRF-based sequential labeling proceeds as follows. First, we applied a Japanese morphological parser (MeCab[2]) to documents and segmented the sentences into tokens with part-of-speech and reading. Then, the relationship between tokens was estimated using CaboCha[3], which is a common implementation of the Japanese dependency parser (Kudo and Matsumoto, 2002). Finally, we extracted the features of the tokens and created models using CRFsuite.

2.2 Basic Features

We used the following features to capture the characteristics of the token: surface, part-of-speech, and dictionary matching. The surface and part-of-speech of the target token were converted into numerical expressions in what is known as one-hot representation: the feature vector has the same length as the size of the vocabulary, and only one dimension is on. The dictionary feature is a binary expression that returns one if a word is in the dictionary and zero if not.

[1] http://www.chokkan.org/software/crfsuite/
[2] http://mecab.googlecode.com/svn/trunk/mecab/doc/
[3] http://code.google.com/p/cabocha/

We prepared ten kinds of dictionaries featuring age expressions, organ names, Japanese era names, family names, time expressions, names of hospital departments, disease names from the Japanese Wikipedia, Chinese characters related to diseases, suspicious expressions, and negative expressions. These dictionaries were created on the basis of the rules explained in Section 3.

To capture the local context of a target token, we combined features of several neighbor tokens. First, we merged the features of five adjacent tokens. Let w_i be the i-th token of the sentence. We concatenated the features of w_{i-2}, w_{i-1}, wi, w_{i+1}, and w_{i+2} and created $w_{[i-2:i+2]}$ to express the i-th node. Second, we concatenated the features of $w_{[i-2:i+2]}$ and w_i^{src} (w_i^{tgt}) to denote source (target) token of w_i.

2.3 Unsupervised Feature Learning

In addition to the basic features, we used clustering-based word features (Turian and Ratinov, et al., 2010) to estimate clusters of words that appear only in test data. These clusters can be learned from unlabeled data by using Brown's algorithm (Brown and deSouza, et al., 1992), which clusters words to maximize the mutual information of bigrams. Brown clustering is a hierarchical clustering algorithm, which means we can choose the granularity of clustering after the learning process has been finished.

We examined two kinds of Brown features: those created from training and test data related to the MedNLP Task (1,000 categories) and those created from the Japanese Wikipedia (100 categories). We decreased the number of categories of the latter because clustering Wikipedia is computationally expensive. The computational time of Brown clustering is $O(VK^2)$, where V denotes the size of vocabularies and K denotes the number of categories.

3 Rule-based Method

In this section, we explain the rule-based method.

3.1 De-identification task

- <a>: age
  - The basic pattern is "d1[歳才台代] (SAI (years old), SAI (years old), DAI (10's, 20's, ..., etc.), DAI (10's, 20's, ..., etc.))", where d1 is a positive integer, and [ABC] refers to A, B, or C.
  - If an age region is followed by specific modifiers "時|頃|[こご]ろ|代|[前後]半|以[上下] (JI (when), KORO (about), DAI, ZENHAN (anterior half), KOUHAN (posterior half), IJOU (over), IKA (under))", that region is expanded to the end of the modifier. A disjunctive expression "aaa|bbb|ccc" means aaa, bbb, or ccc.
  - If an age region is followed by one of interval-markers "から|より|まで|～ (KARA (from), YORI (from), MADE (to))", that region is expanded to the end of the marker.
  - If one age region is followed by another age region directly or with only hyphen-type characters (-ー－—～) between them, the two regions are joined to one.
    - ✧ eg. <a>27 歳 (27 SAI (27 years old))</a>～<a>47 歳 (47 SAI (47 years old))</a>→<a>27 歳～47 歳</a>
- <t>: time
  - The basic pattern of time tags is "d1 年 d2 月 d3 日 d4 時 d5 分 d6 秒 (d1 NEN (year) d2 GATSU (month) d3 NICHI (day) d4 JI (hour) d5 FUN (minute) d6 BYO (second))", where d1 to d6 are non-negative integers. Any partial pattern starting from d1 or d2 or d3 is also eligible.
  - The special numerical pattern d1/d2 (1900 <= d1 <= 2099, 1 <= d2 <= 12) is interpreted as year = d1 and month = d2. In addition, the special numerical pattern "d1/d2 [に|から|より|まで|～] (NI (at), KARA (from), YORI (from), MADE (to))" (1 <= d1 <= 12, 1 <= d2 <= 31) is interpreted as month = d1 and day = d2.
  - Exceptional patterns are: "[同当即翌前][日年月]|翌朝|翌未明|その後 (same year, this year, next morning, ... etc.)".
  - While a time region is preceded by a prefix-type modifier, or followed by a postfix-type modifier, the region is expanded to the beginning or to the tail of the modifiers.
    - ✧ Prefix type modifiers:
      - [翌昨同当本][年月日] (last year, last month, same day, ... etc.)
      - AM/PM type prefix: 午後 (GOGO (PM))| 午前

    (GOZEN (AM) | AM | am | PM | pm
  - ● Ambiguity type prefix: 約 | およそ | ほぼ | 概ね (YAKU (about), OYOSO (about), HOBO (about), OOMUNE (about))
- ✧ Postfix type modifiers:
  - ● [上中下]旬|初め|午[前後]|深夜|早朝|昼|朝方?|夕[方刻]?|[春夏秋冬] (late at night, early in the morning, ...etc)
  - ● Ambiguity type: 頃 | ころ | ごろ | 前後 | 程 | 以[降後前]
- ✧ Intervals (from ~~~ to ~~~)
  - ● <t>...<t> (から|より|まで|〜) → <t>... (から|より|まで|〜) </t>
  - ● <t>aaaaa</t><t>bbbbb</t> → <t>aaaaa bbbbb</t>
  - ● <t>aaaaa</t> [[- ー − — 〜] <t>bbbbb</t> → <t>aaaaa [-ー−—〜] bbbbb </t>

- <h>: hospital
  - ➢ First hospital tags were added by using the below hospital words dictionary composed of seven words, and temporary division tags were added by using the division-word dictionary of 27 words.
    - ✧ Hospital words: 当院|近医|同院|病院|クリニック|総合病院|大学病院 (TOUIN (my/our hospital), KINNI (near hospital), DOUIN (same hospital), BYOUIN (hospital), KURINIKKU (clinic), SOUGOUBYOUIN (general hospital), DAIGAKUBYOUIN (university hospital)
    - ✧ Division words: 外科|眼科|循環器内科|皮膚科|内科 ... etc. (GEKA (surgery), GANKA (ophthalmology), JUNKANKINAIKA (cardiovascular internal medicine), HIFUKA (dermatology), NAIKA (internal medicine) (27 words)
  - ➢ While a hospital region is preceded by any number of division regions, the hospital region is extended to the beginning of the division regions.
    - ✧ <div>内科</div><div>皮膚科</div><h>病院</h> → <h>内科皮膚科病院</h>
  - ➢ If a hospital region is preceded by a sequence of name characters (■), the region is expanded to the beginning of the name sequence.
    - ✧ ■■■<h>皮膚科病院</h> → <h>■■■皮膚科病院</h>
  - ➢ If a division region is preceded by a sequence of name characters, the region is expanded to the beginning of the name sequence, and the tag is changed to a hospital tag.
    - ✧ ■■■<div>内科</div> →<h>■■■内科</h>
  - ➢ As a special case, if a name character sequence is followed by "[をに]?(紹介|緊急)(受診|入院) (SHOKAI (refer), KINKYU (emergency), JUSIN (consult), NYUIN (stay in hospital))", the name character sequence is taken as a hospital region.
    - ✧ ■■■[をに]?(紹介|緊急)(受診|入院) → <h>■■■</h>[をに]?(紹介|緊急)(受診|入院)
- <p>: person name
  - ➢ This tag was skipped.
- <x>: sex
  - ➢ The sex tags were added only by a simple pattern: "男性|女性 (DANSEI (male), JOSEI (female))".

3.2 Complaint and diagnosis task

- All <c> tags of the training data were extracted and a dictionary of complaints was made containing 1,068 words
- The <c> tags were added to the test data by the longest match method using the dictionary. In case of a single character word (咳 and 痰), a tag is added only if both the preceding character and the following character are not Kanji characters.
- If a <c> tag region is followed by the cancelling expressions below, the <c> tag is cancelled.
  - ➢ postfix type cancelling expressions: [歴剤量時室率]|検査|教育|反応|導入|胞診|精査|を?施行|培養|細胞|成分|取り?扱|ガイ[ダド]|分類基準|[^予防]*予?防|[^療]*療法|=[0-9] (history, inspection, prevention, ... etc.)

- <family> tags are added by using the following family-words:
  - 祖父母|兄弟?|姉妹?|[叔祖][父母]|[父母]親?|息子|娘|弟|妹 (SOHUBO (grandparent), KYOUDAI (brother), SHIMAI (sister), CHICHIOYA (father), HAHAOYA (mother), MUSUKO (son), MUSUME (daughter), OTOUTO (younger brother), IMOUTO (younger sister))
- Exception: some of following words are not tagged.
  - 親指|母指|娘細胞 (OYAYUBI (thumb), BOSI (thumb), MUSUMESAIBOU (daughter cell))
- If a <c> tag is preceded by a <family> tag in the same sentence, then "family" modality is added to the <c> tag.
  - <family>祖母</family> ... <c> aaaaa </c> ... <c>bbbbb</c> → <c modality=family> aaaaaa </c> ... <c modality=family> bbbbb </c>
- <negation> tags added to negation words like "ない (NAI (not))" or "ぬ (NU (not)", using Japanese morphological analysis.
- Also negation expressions like "否定的|否定され|(-) (HITEITEKI (negative), HITEISARE (denied))" are tagged with <negation> tag.
- <suspicion>, <recognition> and <improvement> tags are also tagged by pattern matching.
  - suspicion: 疑[いうっ] | 疑わ[しせれ] | うたが[いうっ] | うたがわ[しせれ] | 可能性 | 危険性 | 否定でき<negation> | 考慮され | 考え | 思われ (UTAGAU (to suspect), KANOUSEI (possibility), KIKENSEI (dangerous), KOURYOSARE (considering), KANGAE (think), OMOWARE (appear))
  - recognition: 認め | 診断 | 出現 | 訴え | みとめ (MITOME (recognize), SHINDAN (diagnosis), SHUTSUGEN (appearance), UTTAE (complain), MITOME (recognize))
  - improvement: 改善 | 消失 | 解消 | 離脱 | 軽快 (KAIZEN (improve), SHOUSHITU (disappear), KAISHOU (reverse), RIDATSU (separation), KEIKAI (resolve))
- If an <improvement tag or a <suspicion> tag is directly followed by a <negation> tag, then both tags are cancelled.
  - <improvement>改善</improvement>せ<negation>ず</negation> → 改善せず
  - <suspicion>疑われ</suspicion><negation>ず</negation> → 疑われず
- If a <recognition> tag is directly followed by a <negation> tag, then the recognition tag is cancelled and the negation tag is extended to the beginning of the recognition tag.
  - <recognition>認め</recognition> <negation>ず</negation> → <negation>認めず</negation>
- If a <c> tag is followed by a <negation> tag or <improvement> tag in the same sequence, and if the in-between part (M) does not contain any recognition/suspicion tags, then
  - if no other <c> tag exists in the in-between part M, "negation" modality is added to the <c> tag.
  - if other <c> tags exist in M, and if the in-between parts of <c> tags are composed of the following connecting expressions, then the negation modality is added to the <c> tag.
    - あるいは | または | および | 及び? | 乃至は? | ないしは? | その他の? | など | や | と | 等 (ARUIWA (or), MATAWA (or), OYOBI (or), NAISHIWA (or), SONOHOKANO (other), NADO (and others), YA (or), TO (and), NADO (and others))
- If a <c> tag is followed by a <suspicion> tag, then "suspicion" modality is added under a similar condition as above.

4 Result

4.1 De-identification task

The results of the de-identification task are as follows.

	P	R	F	A
Rule	89.59	91.67	90.62	99.58
ML1	92.42	84.72	88.41	99.49
ML2	91.50	84.72	87.98	99.46

The Rule column shows the results of the rule-based method, and the ML1 and ML2 columns show the results of the machine learning methods. The ML1 is the result with Brown clustering using training and test data of the MedNLP Task. In addition to this, the ML2 is the result using Japanese Wikipedia for Brown clustering.

4.2 Complaint and diagnosis task

The results of complaint and diagnosis task are as follows.

	P	R	F	A
Rule	72.47	58.12	64.50	93.40
ML1	88.98	74.24	80.94	96.08
ML2	88.55	75.32	81.40	96.06

5 Conclusion

For the de-identification task, the rule-based method achieved slightly higher results, while for the complaint and diagnosis task, the machine learning based method had much higher recalls and overall scores. These results suggest that we should use these methods selectively depending on the nature of the information to be extracted, that is to say, whether it can be easily patternized or not.

References

Morita, M., Kano, Y., Ohkuma, T., Miyabe, M., and Aramaki, E. 2013. Overview of the NTCIR-10 MedNLP Task, In *Proceedings of the 10th NTCIR Workshop Meeting on Evaluation of Information Access Technologies.*

Lafferty, J., McCallum, A., and Pereira, F. 2001. Conditional Random Fields: Probabilistic Models for Segmenting and Labeling Sequence Data, In *Proceedings of the 18th International Conference on Machine Learning*, 282-289.

Kudo, T., and Matsumoto, Y. 2002. Japanese Dependency Analysis using Cascaded Chunking, In *Proceedings of the 6th Conference on Natural Language Learning* (COLING 2002 Post-Conference Workshop), 63-69.

Turian, J., Ratinov, L., and Bengio, Y. 2010. Word representations: A simple and general method for semi-supervised learning, In *Proceedings of the 48th Annual Meeting of the Association for Computational Linguistics*, 384-394.

Brown, P. F., deSouza, P. V., Mercer, R. L., Pietra, V. J. D., and Lai, J. C. 1992. Class-based n-gram models of natural language, *Computational Linguistics*, 18:467-479.

The Electronic Health Record as a Clinical Study Information Hub

Naoto Kume
EHR Research Unit, Department of Social Informatics, Graduate School of Informatics, Kyoto University / Kyoto, Japan
kume@kuhp.kyoto-u.ac.jp

Kazuya Okamoto
Division of Medical Information Technology and Administration Planning, Kyoto University Hospital / Kyoto, Japan
kazuya@kuhp.kyoto-u.ac.jp

Tomohiro Kuroda
Division of Medical Information Technology and Administration Planning, Kyoto University Hospital / Kyoto, Japan
tomo@kuhp.kyoto-u.ac.jp

Hiroyuki Yoshihara
EHR Research Unit, Department of Social Informatics, Graduate School of Informatics, Kyoto University / Kyoto, Japan
lob@kuhp.kyoto-u.ac.jp

Abstract

The use of Electronic Health Records (EHRs) is spreading rapidly in several countries. The systems currently used, however, are not designed to permit secondary use of collected data. We present a new design for an EHR system that is capable of connecting information from multiple sites for use in clinical studies by means of an accounting information system and a hospital information system (HIS). This EHR system was designed for healthcare facilities in the Kyoto region. This paper describes how the conventional system can be extended into an EHR system that serves as a clinical information hub.

1 Introduction

EHR is defined by ISO as repository of information regarding the health status of a subject of care, in computer processable form, stored and transmitted securely and accessible by multiple authorized users, having a standardized or commonly agreed logical information model is the support of continuing, efficient and quality integrated health care (ISO 2005). In the United States and New Zealand, an EHR system used by multiple hospitals and clinics in a given region is known as an electronic medical record (EMR) system. Accordingly, this paper distinguishes an EMR system from an EHR system, which is restricted to only one facility.

The original motivation for sharing medical records between facilities was to promote collaboration between clinical facilities. Because medical costs are very high and still rising rapidly worldwide, reducing unnecessary medical tests and duplicate prescriptions is a social necessity. EHR is expected to help accomplish this by improving information sharing. Several countries have established national EHR programs with government support; these include New Zealand, which uses a national healthcare IT plan known as eHealth (National Health Information Technology Board), and Australia, which has established a group (National e-Health Transition Authority) to promote its patient-controlled system. Canada has likewise established a nationwide organization (Canada Health Infoway) to distribute EHR data across all provinces and territories. Finland and Singapore are also promoting the establishment of a national database of medical records.

In most cases, the sharing of medical records is for short-term reference purposes only, and only a document index is exchanged. In Singapore, for instance, secondary use of medical information managed by the public healthcare service (MoHH) is strictly restricted by law. Therefore, the design of Singapore's EHRs is based on document index convergence rather than unified data storage. Since sharing an EHR involves sending and receiving different types of documents, the accumulation of documents can provide a data source extrapolating actual clinical activities. Note that clinical facilities are not guaranteed organization continuity so that the data repository continuity of each facility is also not guaranteed. Therefore, this study focuses on EHR systems with centered repositories storing different types of medical documents. This study aims to provide a design for an EHR that can serve as an information hub and especially as a resource in gathering data for clinical studies.

2 Methods

Within the last two decades, EMR has been installed in most university hospitals in Japan. These installations were performed by different vendors in different hospitals and regions. Therefore, before medical information can be exchanged, users must define a standard format for data exchange. Because there are so many medical document formats in use, no standard format has yet been agreed upon. Instead, every hospital has its own preferred format. EMR systems can be classified based on structure into two types: a centralized database type and an information locator type. The centralized database type receives test results from a lab test branch system and stores as well as the other medical records such as prescriptions, reports and summaries. The information locator type stores only medical records and pointers to the test results and reports; the body of test results and reports are handled by branch systems. From the viewpoint of EHR construction, an EMR of the information locator type is not cost-efficient because it has to collect documents from every single branch system. On the other hand, an EMR of the centralized database type requires only medical records to be converted to a standard exchange format. In order to support analysis of medical records, an EHR aiming to serve as a source for clinical study data should employ the centralized database model; otherwise, the initial cost and update costs could be prohibitive. In fact, although many Japanese hospitals installed EHR systems during the last two decades, when they were subsidized by the government, most of them were abandoned after the subsidy ended.

If medical records are to be used for research in clinical studies, it is necessary to have access to large numbers of cases, especially for studies on rare diseases. The medical records maintained by a single facility are obviously not sufficient for such studies. When the records in an EHR are used, however, it is hard to avoid controversy over document ownership. Sharing policies vary from one facility to the next, so that researchers must pay close attention to the policy of each facility, at least until a standard practice for sharing policy develops. Data sharing and the use of large data sets come with certain disadvantages even though obvious benefits.

Clinical studies are performed in strictly controlled environments so that the relation between treatment and outcome can be firmly established. A comparison between treatment as action and clinical outcome as result is mandatory. General action in a hospital can be defined as orders. Orders are recorded in the form of an order history and a description in the medical records in a HIS. In addition, each action is recorded in an accounting system so that the hospital can claim payment for medical services. Order histories in HISs are not typically standardized, but accounting system records are. In fact, because the accounting report format is determined by the government, all facilities in a country use the same one. Accounting records of hospital orders can therefore be used as an action history, allowing comparison of orders across clinical facilities.

Clinical outcomes are recorded in medical records, lab test results, pathology reports, radiology reports and so on. Those records usually consist of structured and non-structured data. Test results are an example of structured data. Pathology and radiology reports, which are described in natural language, are non-structured. Even in hospitals where all outcomes are managed in a structured data format, the structured data format is different from those used by other hospitals. Therefore, a standard structured data format for each item has to be established and used in multiple facilities before outcomes can be compared. In addition, natural language processing (NLP) is necessary to transform non-structured data into ontology components. Some studies have applied NLP manually to certain document types such as the discharge summary. As this is labor-intensive, however, it is not cost-efficient and therefore not an option for most facilities. Therefore, semi-automatic analysis is required.

Overall, the authors defined the requirements for an EHR intended as a data source for clinical studies as follows:

- Centralized database collecting data capable of traversal query among multi-facility records
- Access control to maintain data-source-facility's sharing policy

- Convergence of accounting information and medical records as action and result
- Automatic data mapping engine using massive medical records

Figure 1 illustrates a proposed design for an EHR generating structured data using its own dataset as well as structure mapping between several facilities' data.

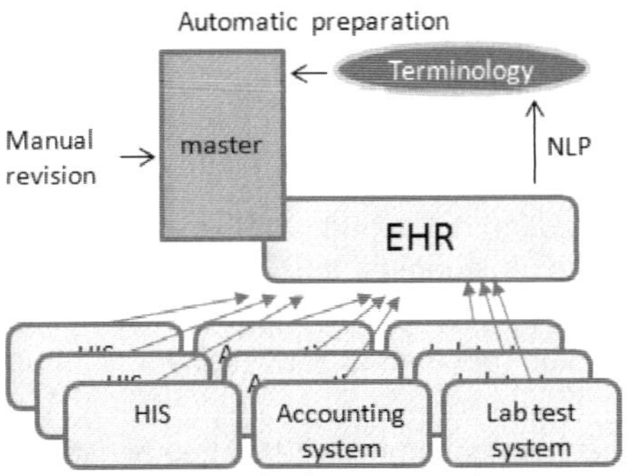

Figure 1. An EHR design promoting convergence of HIS information and accounting information.

If traversal master data is manually maintained, data analysis cannot be completed because of the rate at which new data arrives from other facilities. Therefore, both master data maintenance and medical record exchange requires NLP to achieve semiautomatic data preparation.

So far, there are several obstacles to the realization of this proposed EHR, including data ownership control issues, legal restrictions on data location, the need for NLP technique for automatically generating terminology, the need for a master data definition, and so on. Because we cannot fully implement a real EHR of this kind, therefore, the authors implemented a database to verify how the EHR equipping a centralized database would contribute to a clinical study. Specifically, the authors implemented a clinical study database composed of datasets from several university hospitals. The centralized database constitutes a traversal search environment for accounting information and test results.

There are six requirements of an EHR database suitable for clinical use: available data range, access control, sharing control, search query performance, usability, and database management policy agreement. Available data range depends on the NLP technique as well as the reachable dataset in the HIS. Data export from the HIS to the EHR depends on the conventional hospital setup and policy. Exported data is aligned in a semi-structured format such as XML. Access control must be extended to researchers as well as medical caregivers and patients. Also, for privacy reasons, results should consist of overviews and abstractive information instead of patient-specific information. Sharing control should be given to each participating hospital's administration; otherwise, the hospital's internal council will hardly be convinced. The database should also be fast enough to allow for a traversal search of multiple datasets that are each constantly growing. To enable such fast searching, each dataset should be optimized for a search query created by a manager who is capable of setting up each researcher's required database query. Because it is too hard for researchers to correctly understand this data retrieved from multiple datasets, a search query manager should be assigned at the datacenter. Finally, an audit council should be organized to reach agreement on any database management issues.

Because of these restrictions, it would be difficult to implement the EHR as designed from scratch. Therefore, as a proof-of-concept, the authors implemented a system that meets the above requirements but is based on a currently available technique and dataset. There are two convergent datasets, a set of accounting information and a set of test results, as action and result information. Four university hospitals contributed to the convergent datasets: Kyoto University Hospital, Chiba University Hospital, Osaka University Hospital, and Miyazaki University Hospital. Although the use of these datasets was approved, the physical setup of a unified database was not allowed. Therefore, the authors installed a virtually centralized database based on a database (Cache). Figure 2 illustrates a database. Each hospital has own dataset in the facility. A virtual datacenter is allowed to access all database by a search query.

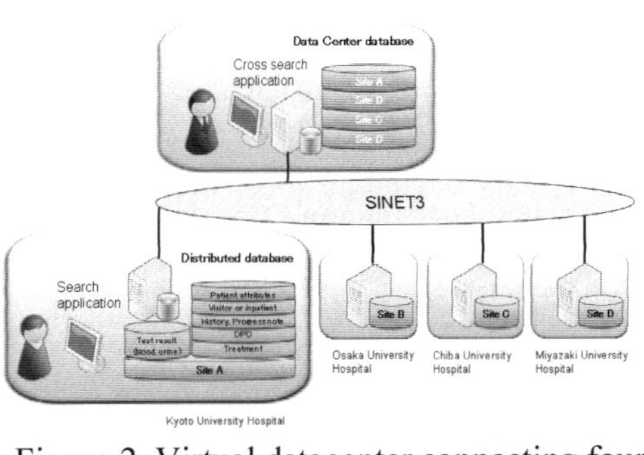

Figure 2. Virtual datacenter connecting four university hospital databases.

A traversal search query is implemented at the datacenter, then equally distributed to all facilities so that each facility can see the search results. Here, if a facility declines to share a search result, an administrator at a local site can stop sharing that result. Also, each search query is discussed and agreed upon by a council consisting of four university administrators beforehand.

So far, there is no proper standard format for test results and no semi-automatic technique for master preparation, so the authors manually prepared a unified master based on JLAC10 and composed of four university test result masters. The accounting information master used here is well established because the format for medical treatment fee claims is almost universal in all hospitals in Japan.

3 Results

The authors carried out five studies using the database. A query manager created a traversal search query by making an optimal search query of each site database. Because the unified test result master was manually maintained, only 300 test result items were available for query, even though over 3000 items are available in each site. The script list as a researcher would see it is presented as Figure 3. The researcher's request was analyzed by the manager and translated to a database query beforehand. The authors proceeded with five traversal queries as follows: Zyvox-treated patients, nicotine addiction treatment patients, teicoplanin-treated patients, bortezomib-treated patients, and influenza patients. Here, the definition of patient is different between facilities because of differences in employed drugs, drug names and applied disease names. Therefore, the query manager was required to find proper combinations of those parameters optimized to each facility's database.

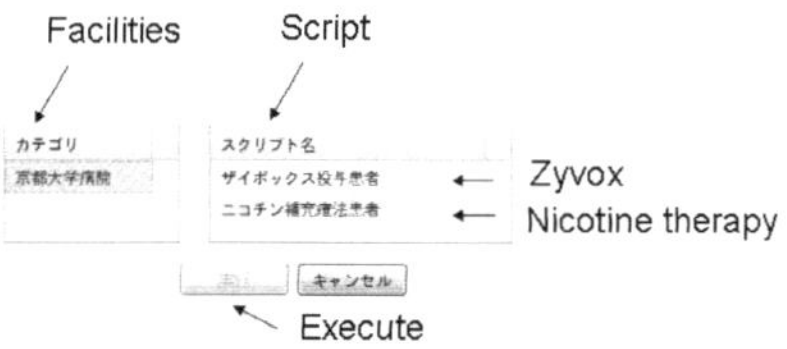

Figure 3. User interface of traversal search query script.

Figure 4 illustrates the result of a search of patient findings. The result can be evaluated by a site administrator before it is shared. If the result is allowed to be shared, the list is accumulated to a statistical result.

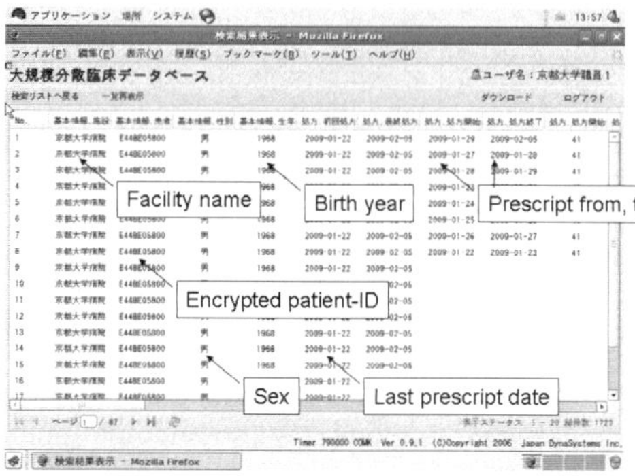

Figure 4. Traversal search result list for a site administrator.

The virtually centralized database, handling 180 million records including data from 2009 and 2010, completed each query in less than 10 seconds. It was concluded that this level of information processing performance is enough to satisfy usability.

The results show that the number of patients can be compared between hospitals. In other words, a centralized database like this can be beneficial for case finding, especially for finding rare diseases and common treatment procedures.

General patient information such as height, weight, medical questionnaire responses, disease history, vital information, and contraindicated medicines are necessary and must be added to the database, according to a pharmacoepidemiologist.

4 Discussion

Clinical studies require strict control of data to verify the results. On the other hand, epidemiology requires a massive dataset to analyze general information statistically, even if the accuracy of the dataset items is not verified. Therefore, the proposed EHR would contribute to epidemiology as well as case finding in clinical studies. The data sharing policy should begin with the patient's ownership of his or her own data, for no unified database can be achieved based on caregiver's ownership. Also, a semi-automatic data alignment technique to maintain master data and analyze unstructured documents is necessary. NLP would be convoluted to the data cleansing cycle.

5 Conclusion

This paper proposed an EHR designed to serve as an information hub for clinical studies. A centralized EHR database was defined to achieve traversal medical record search of multiple facilities. The authors implemented a virtually centralized database as a proof-of-concept. The database contained accounting information and test results. Five case studies were performed to find patients from multiple facilities. The authors concluded that the database cannot be used directly in clinical studies but is beneficial in case finding as well as epidemiologic analysis.

References

Canada Health Infoway, https://www.infoway-inforoute.ca/ (Last access: Jul 20, 2013).

Cache, InterSystems Corporation, http://www.intersystems.com/cache/index.html (Last access: Jul 20, 2013).

ISO 2005, "Health informatics — Electronic health record — Definition, scope and context", TECHNICAL REPORT ISO/TR 20514:2005(E), p. 2, 2005.

MoH Holdings in Singapore, http://www.mohh.com.sg/about_mohh.html (Last access: Jul 20, 2013).

National Health Information Technology Board, http://www.ithealthboard.health.nz/ (Last access: Jul 20, 2013).

National E-Health Transition Authority, http://www.nehta.gov.au/ (Last access: Jul 20, 2013) .

Author Index

Aramaki, Eiji, 1, 7, 33

Fujita, Nobuyuki, 30

Hasida, Koiti, 33
Higashiyama, Shohei, 14

Ikeda, Yukichi, 22
Imaichi, Osamu, 38
Isotani, Ryosuke, 22

Kaneko, Shuj, 30
Kume, Naoto, 43
Kuroda, Tomohiro, 43

Masuichi, Hiroshi, 1
Matsubara, Yusuke, 33
Matsumoto, Yuji, 22
Miura, Yasuhide, 1
Miyabe, Mai, 33

Nakamura, Satoshi, 22
Neubig, Graham, 22
Niwa, Yoshiki, 38

Ohe, Kazuhiko, 1
Ohkuma, Tomoko, 1
Ohtake, Hiroshi, 30
Okamoto, Kazuya, 43
Okumura, Takashi, 7

Sakti, Sakriani, 22
Seki, Kazuhiro, 14

Takaai, Motoyuki, 33
Tam, Wailok, 33
Tateisi, Yuka, 7
Toda, Tomoki, 22

Uehara, Kuniaki, 14
Uozaki, Hirosi, 33

Yamada Shinohara, Emiko, 1
Yanase, Toshihiko, 38
Yoshihara, Hiroyuki, 43